# The Capitol Reef Reader

The National Park Readers
Lance Newman and
David Stanley,
series editors

# THE CAPITOL REEF READER

EDITED BY STEPHEN TRIMBLE

THE UNIVERSITY OF UTAH PRESS
*Salt Lake City*

Frontispiece: Cars in Capitol Gorge. Arthur Inglesby photo collection, P0899 n144, Special Collections, J. Willard Marriott Library, University of Utah, Salt Lake City. Used with permission.



The Defiance House Man colophon is a registered trademark of The University of Utah Press. It is based on a four-foot-tall Ancient Puebloan pictograph (late PIII) near Glen Canyon, Utah.

LIBRARY OF CONGRESS CATALOGING-IN-PUBLICATION DATA

Names: Trimble, Stephen, 1950– editor.
Title: The Capitol Reef reader / edited by Stephen Trimble.
Description: Salt Lake City : The University of Utah Press, [2019] | Includes bibliographical references and index. |
Identifiers: LCCN 2018047884 (print) | LCCN 2018049960 (ebook) | ISBN 9781607816836 () | ISBN 9781607816829 (pbk. : alk. paper)
Subjects: LCSH: Capitol Reef National Park (Utah)--Description and travel. | Capitol Reef National Park (Utah)—History. | Natural history--Utah—Capitol Reef National Park.
Classification: LCC F832.W27 (ebook) | LCC F832.W27 C37 2019 (print) | DDC 979.2/54--dc23
LC record available at https://lccn.loc.gov/2018047884

Errata and further information on this and other titles available online at UofUpress.com

Printed and bound in the United States of America.

The Nature Conservancy of Utah has generously supported the publication of this volume.

*For Joanne—*

*my partner in adventure*

*When I made my first trip to Capitol Reef, it was those spooky, unimaginable shapes of stone that caught and held my attention.... Domes, pinnacles, turrets, spires, arches and bridges, pits, pockets, potholes, plunge pools, goblins and hobgoblins, cracks and crevasses and canyons, buttes, cathedrals, caves, grottoes, alcoves, humps, hills, holes and hollows, slick walls and sheer cliffs—there is, as far as I know, no other place on earth where time and weather have created so curious and so intriguing an assemblage of landforms....*

*To get the feel of this country, to taste its atmosphere and sample its spirit, you must leave your car and go up into the hills of golden stone, up into that never-never land of petrified cities, secret waterpockets crawling with fairy shrimp and mosquito larvae, and into those mysterious little canyons which lead nowhere, where no human has ever gone before, for all you know. You must follow those dim trails which take you out of civilization's web, beyond all that is safe and familiar, out of sight, out of mind, out of reach.*

—Edward Abbey, "A Thirst for the Desert," 1973

## CONTENTS

Introduction: Paradise & Slickrock by Stephen Trimble 1

**Part I. The View from Boulder Mountain: The Geography of Hope** 17

1 Geology of the High Plateaus 19
Clarence Dutton

2 Wilderness Letter 22
Wallace Stegner

3 Into the Escalante 24
Linda Elizabeth Peterson

Portfolio: The Bright Edge 27
Photographs by Stephen Trimble

**Part II. Native Heritage: Culture & Memory** 35

4 Dwellers of the Rainbow 39
Rose Houk

5 Fremont Places, Fremont Life, Fremont Place 46
Steven R. Simms

6 Fulfilling Destinies, Sustaining Lives 51
Rosemary Sucec

7 Seeing Is Believing: The Pectol Shields 56
Robert McPherson

**Part III. Exploration: Walking on the Flinty Mountains** 67

8 Incidents of Travel and Adventures in the Far West 71
Solomon Nunes Carvalho

9 Sights Once Seen: Daguerreotyping Fremont's Last Expedition through the Rockies 79
Robert Shlaer

10 Photographed All the Best Scenery 85
Jack Hillers

11 Gilbert's Notebooks 90
Charles B. Hunt

12 In Cathedral Valley 94
George C. Fraser

13 One of the Great Trips of the World 99
John A. Widtsoe

**Part IV. Canyon Country: Earth Unmasked** 107

14 An Introduction to the Park 109
Ward J. Roylance

15 Come on In 112
Edward Abbey

16 Impracticable Ridges 117
Charles Bowden

17 The Neverlasting Hills 122
Michael Collier

18 A Field Guide to Brazen Harlotry 128
Ellen Meloy

19 Vignettes from "The Near-Sighted Naturalist" 135
Ann Zwinger

20 Feathered Cultivators 139
Ronald M. Lanner

**Part V. Home: Listening to the Desert** 145

21 White House on the River 147
Ruby Noyes Tippets

22 Pleasant Creek History 156
Lenard E. Brown

23 Two Marshalls 159
Sidney A. Hanks and Ephraim K. Hanks Jr.

24 Too Lonesome Down There 167
Billie Bullard

25 Eating the View 173
Lurt Knee

26 Landing under a Sleeping Rainbow 178
Chip Ward

27 The New Face of Nature 186
Utah Valley University Capitol Reef Field Station Students

28 The O-Bar Drive 191
Ray Conrad

**Part VI. Fruita: Verdant Village, Sweet Sanctuary** 197

29 Pick, Pick Apples 199
Kate MacLeod

30 The Junction 201
Franklin Wheeler Young

31 Echoes of Childhood in Old Fruita 202
Clay M. Robinson

32 Poverty Flats & Paradise 208
George Davidson

33 Jeweled Jars of Memory 215
Jen Jackson Quintano

34 High Plateaus 219
Wallace Stegner

35 Red Rocks and Eccentrics 225
Renny Russell

36 Charles Kelly 232
Gary Topping

37 The Forgotten Park: Charles Kelly 234
Jonathan Thow

Part VII. National Vision: Crafting Wayne Wonderland 239

38 Fix the Roads, the Tourists Are Coming 241
Jared Farmer

39 The Forgotten Park: Uranium Mining 247
Jonathan Thow

40 From Barrier to Crossroads: The Genesis of Expansion, 1968 254
Bradford Frye

41 Putting Together the National Park 265
Lurt Knee

42 Right-of-Way 267
Jared Farmer

Part VIII. Refuge: In the Wild Heart of the Fold 275

43 Capitol Reef's Most Famous Photograph: Minor White's *Moenkopi Strata, 1962* 277
James Swensen

44 Capitol Reef, Illustrated 284
Stephen Trimble

45 Shouting at the Sky 292
Gary Ferguson

46 So What If We All Want Redemption 298
Ann Whittaker

47 Landscape of Desire 300
Greg Gordon

48 Rock Glow, Sky Shine: The Spirit of Capitol Reef 306
Stephen Trimble

49 October Journey 310
Rose Houk

50 The Trouble with Cell Phones 316
Craig Childs

51 Discover Your Own Private Paradise 318
Steve Howe

52 Spoked Dreams: An Odyssey by Bicycle and Mind 323
Charles Riddel

**Part IX. Practicable Ridges: Nurturing the Future** 329

53 Bargaining for Eden 331
Stephen Trimble

Capitol Reef Timeline 339

Further Reading 341

Acknowledgments 343

Sources and Chapter Notes 345

Looking upstream along the Fremont River, through the Blue Gate toward the Waterpocket Fold and Boulder Mountain, from the rim of North Caineville Mesa, 2005.

# INTRODUCTION

## PARADISE & SLICKROCK

"Rock. No other word can precede it."

I began my 1978 interpretive booklet about Capitol Reef with these words, and I begin my introduction to *The Capitol Reef Reader* with the same words. Every traveler who passes through the park encounters a panorama ruled by geology, where rabbitbrush, raven, and rock art are accents, where the layers of earth history overwhelm every other element of the scenery.

I felt this for the first time when, in 1975, at twenty-four, I found myself headed into Capitol Reef on Utah's Highway 24 to begin my season as a park ranger–naturalist. The redefinition of "Capitol Reef" was new, too—after its sixfold enlargement as a national monument in 1969 and subsequent designation as a national park in 1971.

I came from the east, through the austere outcrops that locals call the Blue Gate, one of two "gates" that frame Capitol Reef country. The highway from Hanksville approaches the park between these fluted skirts of blue-gray shale that drape the slopes of surrounding mesas. From the opposite side, on the west, it's the Red Gate between Bicknell and Torrey that serves as portal, a first flare of sandstone cliffs where the outrageous colors of canyon country announce themselves, drawing travelers deeper into the redrock wilderness, downstream toward the Colorado River.

Each of Utah's five national parks—Arches, Canyonlands, Capitol Reef, Bryce Canyon, and Zion—defines a special place on the Colorado Plateau, the vast sweep of flat-lying rocks surrounding the Four Corners and eroded into a maze of canyons by the Colorado River. I'd spent time exploring these four other parks, but Capitol Reef was virgin territory for me. I had done no more than drive the highway that bisected Utah's least-known park at its narrow waist, once with my parents on vacation from our home in Colorado and once on a road trip with my college buddies. Now I was moving here to live for a season, with all my belongings crammed into my pinkish-tan 1962 Dodge Dart.

I kept heading west, swinging around the curves and against the grain of the Waterpocket Fold, the tipped-back wrinkle of rock layers that runs for a hundred miles across southern Utah. Each mile took me into new strata, older rocks, each with its own textural personality and mineral-stained flamboyance. When I reached what I soon would learn was the bentonite clay of the Morrison Formation, I remember calling out in wonderment, "My god, those rocks are *purple*!"

I knew enough about these rocks to know that they recorded what geologists call "deep time," layer after layer deposited in environments so different from today. Oceans, deserts, dinosaur-filled swamps. It was as if all the earth's history had been distilled into a single rock column in this one spot.

My highway followed the Fremont River, a ribbon of life-affirming green in a daunting expanse of stone. Cottonwoods defined this verdant corridor, as both river and road threaded the canyon between golden domes of Navajo Sandstone. And then the Fremont canyon widened. The fractured red wall of the Wingate Sandstone lifted and pulled away from the river, creating a tiny, isolated pocket of irrigable fields with a grid of fruit trees between river and cliff. A pioneer one-room school, a barn, and an old Mormon farmhouse punctuated all that green like a movie set.

This was the village of Fruita, a cherished oasis in the labyrinth of canyons of the Colorado Plateau, the single outpost of settlement within the small original Capitol Reef National Monument created in 1937. This would be my home for the next seven months.

In the more than forty years since I arrived at that seasonal ranger job at Capitol Reef, I've lived all over the Four Corners states, but this many-layered little haven where water meets cliff is still the heart of my spiritual home.

My job as a ranger consisted of paying attention—and sharing my newly acquired stories, photographs, and knowledge with park visitors. I was fresh to this place, fresh to teaching, fresh to professional responsibility. I drew my

The historic Pendleton-Gifford barn, Fruita, 1996.

sustenance from the land, from hiking the backcountry, and from reading the writers who had captured the Plateau with their words, Wallace Stegner and John Wesley Powell and Ed Abbey, Ann Zwinger and Willa Cather—and then doing my best to draw in park visitors as new to the country as I was and to galvanize them with my enthusiasms. I was in love with the place. I was in love with *learning* about the place. And I wanted every traveler to fall in love as I had, to leave Capitol Reef bubbling with exhilaration about the stunning union here of water and rock, time and twisted juniper.

And so here I am, all these years later, sifting through decades of great writing about Capitol Reef to share with readers and park visitors once again my love for this place. *The Capitol Reef Reader* gathers 160 years' worth of words that explore the spirit of the park and its surrounding landscape in personal narratives, philosophical riffs, and historic and scientific records. I've chosen these pieces for the character of their storytelling and language while I've looked for an array of writing that reveals Capitol Reef in all its layers, from geology to history, from Native peoples to twenty-first-century canyoneers. I found pieces that are a pleasure to read and authors who tell

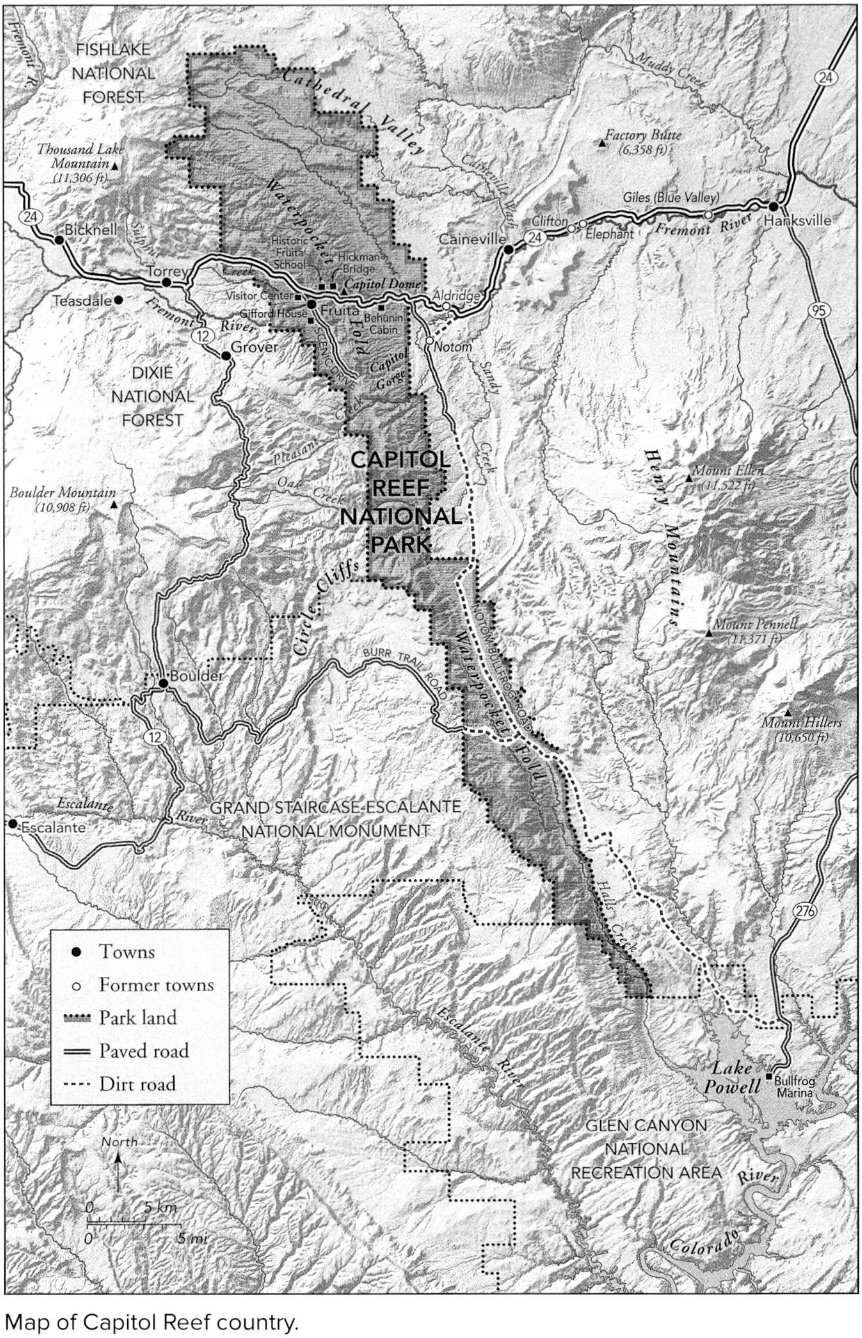


Map of Capitol Reef country.

us better than anyone else about some aspect of Capitol Reef. For more specialized information, see the reading list at the end of this book.

In this reader, I define Capitol Reef not by the right-angled zigzags on a map but by geography. The Waterpocket Fold is the essential landform—whether you are on the ground or looking at Utah from space. This narrow, sharp flex in the flat-lying rocks of the Colorado Plateau begins at the north at Thousand Lake Mountain, arcs across the east-west course of the Fremont River, and runs south all the way to the Colorado River.

To build your mental map of Capitol Reef country, begin with the view eastward from Boulder Mountain, as did so many nineteenth-century explorers perched at that same 11,000-foot-high rim—a vertical mile above Fruita. Boulder Mountain is the easternmost rampart of the High Plateaus, the chain of great flat-topped mountains that separate the inner canyonlands, carved by the Colorado River and its network of tributaries, from the Great Basin to the west. Below your roost on Boulder Mountain, the sandstone domes and reefs of the Waterpocket Fold run in the middle distance to left and right, and they keep running, clear out of sight.

In that maze of canyons lie a handful of gorges that provide passage through the Reef and an even smaller number of permanent streams.

Off to the right, the Fold runs behind the Circle Cliffs and the canyons of the Escalante River—both protected by the enormous Grand Staircase–Escalante National Monument proclaimed in 1996. At the far left, slickrock yields to softer desert landscapes along the northeastern base of the Fold. Here, open badlands and free-standing stone temples in Cathedral Valley, Hartnet Draw, and South Desert give the isolated north end of the park a starkness unlike any other terrain in Capitol Reef.

Breaking the horizon beyond the Waterpocket Fold rise the five island summits of the Henry Mountains, east of the park but dominating the view from every high point in Capitol Reef.

Descend from Boulder Mountain along Highway 12. The big views continue northward to Torrey, gateway community for Capitol Reef National Park. You've arrived in Wayne County and its string of villages along the Fremont River. You've landed in the heart of the region's history and culture, reaching back to prehistory.

Upper Cathedral Valley, Capitol Reef, 2016.

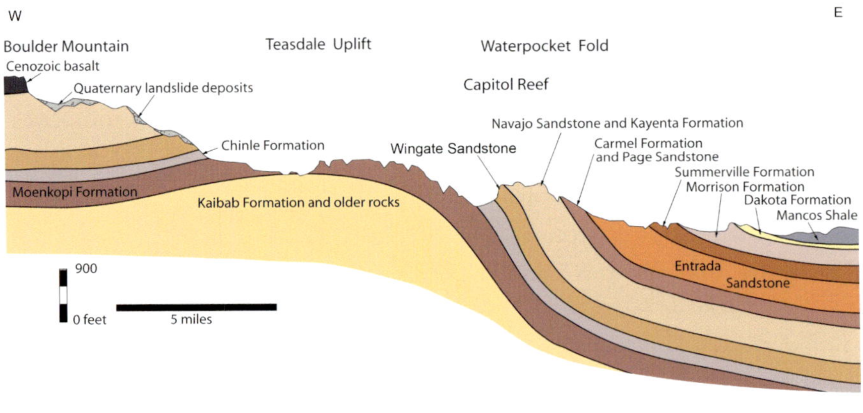


Geologic cross section of the Waterpocket Fold and Capitol Reef country. Courtesy Ron Blakey, Colorado Plateau Geosystems/Deep Time Maps.

Humans have lived in Capitol Reef for at least 12,000 years. We don't think much about the first 11,800 of those years when we talk about "pioneers." But the park's first residents left their mark—most easily seen in the petroglyphs and pictographs sheltered by cliffs along the Fold. Walk nearly any clearing in the park and keep an eye out for chippings and projectile points. A given lithic scatter may not mark a particular moment and culture but rather multiple waves of people inhabiting the congenial spots for hunting and farming, for farmsteads and hamlets, over millennia.

Capitol Reef's river gives us the name for one of these great prehistoric cultures living across a vast swath of Utah, the Fremont. In this reader, writer Rose Houk and archaeologist Steven Simms introduce us to the complex relationship that the Fremont people established with this land. By the time the first soldiers and scientists from the United States encountered Capitol Reef, the Fremont and their neighbors to the south, the Ancestral Puebloans, had given way to Ute, Paiute, and Navajo people.

National Park Service anthropologist Rosemary Sucec documented Native connections to Capitol Reef in the stories she recorded in collaboration with the tribes in the 1980s, and I include her summary of this work here.

The two richest veins of literature to be mined for this reader come from the rim of Boulder Mountain and from the village of Fruita. I begin the reader looking out over Capitol Reef from Boulder Mountain with Clarence Dutton, Wallace Stegner, and Linda Elizabeth Peterson, astonished and awed. Then we move into the canyons with the Mormon pioneers, sharing a bit of their boulder-strewn lives.

The nineteenth-century government surveys came first: John Charles Fremont, mostly lost (a harrowing trip described by the expedition photographer Solomon Carvalho and retraced by contemporary photographer Robert Shlaer), then Lieutenant George Wheeler, and finally the Powell Survey. These explorer-scientists made the region their own. When geologist Charles B. Hunt came to the Henry Mountains fifty years later, in the 1930s, he learned as much from the journals of his long-gone mentor, Grove Karl Gilbert of the Powell survey, as he did from the mountains themselves.

By the early 1900s, you could hire a guide like David Rust. Raised mostly in Wayne County, for decades he oriented newly converted pilgrims, shared his favorite "lookoffs" from the rims of the High Plateaus, and camped his way through the slickrock and sage. One of his customers, George Fraser, takes us to Cathedral Valley in 1915.

Geologists continue to make pilgrimages to the Waterpocket Fold and the Henry Mountains—two classic field localities known worldwide for their defining structures. Nearly every writer tackling Capitol Reef geology finds a relevant quote in the writing of Powell's men, Dutton and Gilbert.

In the 1870s, at the same time that John Wesley Powell and his company of geologists and topographers began to define and explore the Colorado Plateau, Mormon pioneers came to Wayne County. These settlers first chose the upper Fremont Valley, grazing their cattle in the most extensive meadows found in the 2,500-square-mile county.

As families grew and new people joined communities, by the 1880s the Mormon colonists had moved through the Red Gate to plant new settlements at Teasdale, Torrey, and Fruita, following the Fremont River along the line of cliffs that formed a barrier they compared to an ocean reef. They noted the domes of golden Navajo Sandstone that rose with the commanding presence of the US Capitol, and Capitol Reef had its name.

These decades at the end of the nineteenth century brought the first permanent residents since Fremont times to the land within the modern park's boundaries. First, the secluded Floral Ranch on Pleasant Creek, established by Eph Hanks in 1881, then Fruita in 1882. Within twenty years, Fruita was a stable home for a handful of families whose orchards provided a steady annual harvest of fruit for all of Wayne County. Within another twenty years, local folks realized that their home territory was worthy of protection and promotion, and the campaign for a "Wayne Wonderland National Park" had begun.

The park's rare permanent streams drew anyone looking for a home. In the reader, we spend time with four generations of families who lived deep within the park along Pleasant Creek. We also hear from three astounded international students on their first visits to the university field station that now occupies the Pleasant Creek ranch site.

Elijah Cutler Behunin established beachheads of domesticity in the Fremont canyon. He applied his well-intentioned industriousness to channelizing the river (recounted by his granddaughter Ruby Tippets) and unwittingly contributed to the devastating floods that wiped out most of these settlements by the early twentieth century.

The families clustered at Fruita were more successful, and their recollections of their home in the canyons give us the most intimate record of life in Capitol Reef. I experienced a bit of that life when I lived in park headquarters in old Fruita. I watched mule deer come off the mesa each night to bed down in the historic orchards. I reveled in my share of peaches and plums and apricots and apples from the heritage trees. I matched the pace of my life to the changing light from sunrise to sunset on the Big Red Cliff of Wingate Sandstone that rose above our homes.

For the reader, I've had to make tough choices from the wide array of Fruita reminiscences. These pieces capture the charm and challenge of living a hardscrabble life here. Composer Kate MacLeod's pithy lyrics sum up this tension between "the middle of nowhere" and "home"—Eden, God's country, the center of the universe.

From the 1930s to the 1960s, outsiders joined Fruita's deep-rooted Mormon families, and the eccentricities of these newcomers hiding out in remote rural Utah made for good company and promising copy for visiting writers. The first superintendent of Capitol Reef, Charles Kelly, was the crankiest of these originals, and his story receives extra attention.

Mule deer on the mesa above Fruita, 1975.

Capitol Reef superintendents have considerably better relations with Wayne County today than they did in Charlie Kelly's day—with park managers often making their homes in local communities. Even so, Capitol Reef continues to lie at the center of an ongoing give-and-take about the highest and best use of this redrock wilderness, an emotional dialogue that long ago sparked what scholar Charles Wilkinson calls "Fire on the Plateau."

The origins of the park date to the 1920s, when local boosters Ephraim Pectol and Joseph Hickman began campaigning for a park designation that would bring good roads, publicity, and growth to out-of-the-way Wayne County. In the 1930s, conservationists realized that the vast roadless areas of southern Utah were precious and in need of preservation. FDR's Interior Department under Harold Ickes proposed an Escalante National Monument that likely would have saved Glen Canyon on the Colorado from the dam that inundated it thirty years later. Ickes's potential preserve embraced canyon country that eventually would become portions of Grand Staircase–Escalante National Monument, Capitol Reef National Park, Glen Canyon National Recreation Area, Canyonlands National Park, and the 2016 Bears Ears National Monument—as well as the still-unprotected Henry Mountains and Dirty Devil canyons.

Ickes's proposal—though never enacted—laid out battle lines we've worked within for the subsequent eight decades. Cattlemen, miners, and politicians concerned about private-property rights have been trying to move their positions forward in the face of fierce defense from those who see these public lands as one-of-a-kind redrock treasures of biodiversity, cultural history, wildness, and refuge. Tourism creates an intersection, common ground—a less consumptive use capable of generating livelihoods for many. Every position has its trade-offs.

Historians Jared Farmer, Jonathan Thow, and Bradford Frye narrate these conflicts, from the concerns of Wayne County cattlemen in the 1930s to the strike-it-rich dreams of uranium miners in the 1950s to local outrage over the expansion of Capitol Reef National Monument in the 1960s. Skirmishes flared again a generation later when county officials insisted on paving the Burr Trail, and backcountry devotees fought to preserve the solitary feeling of driving that dirt road.

Some visitors to Capitol Reef come to create art. James Swensen illuminates the backstory for the great Minor White photograph of Moenkopi Sandstone that turns up in so many photography texts. I introduce other signature artists of Capitol Reef, both photographers and painters.

Visitors also come to do science, to seek adventure, to observe and enjoy plants and critters, to heal from psychological wounds, and to hike, bike, and canyoneer (a notch more technical than hiking, descending the canyons using ropes and rappel anchors). These visits generate a steady stream of writing on "the Plateau Province."

Naturalist writers walk through these redrock canyons with endless curiosity, seeking connections between their observations of plants and animals and everything else they know—and striving for the words to articulate these relationships. They define themselves primarily as writers, reticent about their skills as naturalists, careful to credit the "real" scientists.

Edward Abbey wrote *Desert Solitaire* about Arches National Park, but his wanderings in the canyons made him the voice of the entire Plateau. Ann Zwinger came to the Colorado Plateau as an artist and art historian, and her descriptions of the colors of tadpoles and aspen bark come right out of her box of watercolors. Ellen Meloy sees the sensual world of canyon country wildflowers—"slickrotica"—more as a writer than as a botanist, but as she explores her emotional bonds with this land, we soak up a lot of thoroughly researched natural history.

For park interpreters, content informs wonder. Rose Houk and I, each of us former park rangers, reveal this fondness for facts in our journeys down Halls Creek.

*The Capitol Reef Reader* ends with a chapter from my book *Bargaining for Eden*, in which I try to unravel the threads that make contemporary Wayne County a place of cherished retreat and a stage for a fundamental clash in values.

As visitation even here in Utah's "forgotten park" passes a million people a year, the dream of 1920s boosters seems realized beyond anyone's imaginings. But how do we make sure all these visitors know to avoid stepping on fragile biological soil crust? How do we make sure they understand enough about the legacy of Native peoples to caress chippings and arrowheads with

Crossbedding, Navajo Sandstone, top of the Waterpocket Fold above Pleasant Creek, 2014.

Fremont River canyon and Highway 24 at Fruita, 2016.

reverence and then return these precious bits of stone to the sand, to observe rock art but never touch? How do we make sure every visitor knows that adding graffiti to the petroglyph panels constitutes an act of vandalism?

I pondered the ironies of twenty-first-century Capitol Reef on a recent journey with my family to the rim of Hall Mesa in the southern backcountry of the park.

We drove our Subaru south from Highway 24 onto Notom Road. Pavement gave way to dirt. We wound our way through fields of late-summer sunflowers near Bitter Creek Divide, passed the Burr Trail and The Post, and took four more turns onto ever-rougher roads.

We were sixty miles from the Capitol Reef Visitor Center. We had seen no other vehicles. We parked when the road got too rough and walked for a half hour to the rim above Halls Creek Narrows, where a peregrine falcon took off from an aerie on a thousand-foot cliff and sailed off into space, screaming at us in warning.

What place could be wilder? Ninety-three percent of the park falls in officially mapped "primitive" ("essentially wild and undeveloped") and "semi-primitive" (more frequent roads or evidence of grazing) zones. The Park Service manages this backcountry as wilderness. Capitol Reef has world-class dark-night skies. What place could be more remote?

We settled on a rocky perch for lunch—and discovered that in this place where so few venture, we had cell phone service. On a whim, we called family in San Diego. "Guess where *we* are???"

Our next question could have been: In twenty-first-century Capitol Reef, what defines "remote"? Where, truly, are we? What decisions will we make to plan effectively for a future Capitol Reef linked to issues far beyond the park—the looming climate catastrophe, deteriorating regional air quality, the conflict over President Donald Trump's decision to drastically reduce the size of nearby Grand Staircase–Escalante and Bears Ears National Monuments, and a sensible accommodation to a Utah population that will double by 2050? We must consider preservation and enjoyment, our need for refuge and solitude, the conservation imperative, working always within the minuscule National Park Service budget.

We will make these decisions most wisely if we know everything we can about Capitol Reef. Ideas, knowledge, and resilience lie within the rootedness of Native peoples, in the affectionate memories of longtime locals who love their home, in the sharp observations of naturalists and scientists, in the professional expertise of park staff, in the creative responses to the landscape by writers, photographers, and artists, in the backcountry puzzles worked out by hikers.

These people know Capitol Reef. Their stories await, spilling from this book. And so, as Ed Abbey says, "Come on in and see for yourself."

Meeks Mesa, 2014.

Looking across Happy Valley toward South Desert and the San Rafael Swell, from Boulder Mountain, 1997.

# PART I

## THE VIEW FROM BOULDER MOUNTAIN

### The Geography of Hope

Capitol Reef's written record opens with the big views off the east rim of Boulder Mountain, vistas so dramatic, so vast, they command the attention of every writer who stands here.

Franklin Wooley comes first, when he describes this "dazzling" scene in his journal for the Andrus Expedition of 1866. Captain James Andrus led a company of the Utah Territorial Militia here, on patrol after the Ute and Southern Paiute people had unsuccessfully fought the theft of their southern Utah homelands in what historians call the Black Hawk War of 1866.

Wooley, the expedition chronicler, looked down from "the black volcanic precipice," stunned by the scale of his view. "Stretching away as far as the Eye can see a naked barren plain of red and white Sandstone crossed in all directions by innumerable gorges.... Occasional high buttes rising above the general level, the country gradually rising up to the ridges marking the 'breakers' or rocky bluffs of the larger streams. The Sun shining down on this vast red plain almost dazzled our eyes by the reflection as it was thrown back from the fiery surface."

Pilgrims keep coming, looking out over Capitol Reef while driving today's Utah Highway 12 between Torrey and Boulder and then, after devouring the views, remembering to breathe.

Winter view toward the Waterpocket Fold and Henry Mountains, from Boulder Mountain, 2018.

## GEOLOGY OF THE HIGH PLATEAUS (1880)

### CLARENCE E. DUTTON (1841–1912)

*Within five years of the 1866 Andrus expedition, John Wesley Powell had sent his fabled band of mapmakers and geologists to the heights of Boulder Mountain. The scientists worked their way east toward the Waterpocket Fold and Henry Mountains—Powell's designation for this last unnamed range in the lower forty-eight. Powell's man Clarence Dutton, a Yale-educated US Army officer with a literary flair, took on the task of deciphering the geology of the High Plateaus—including the Aquarius Plateau, which incorporates Boulder Mountain. Dutton's 1880 Powell Survey monograph,* Report on the Geology of the High Plateaus of Utah, *includes his contemplative scan from the 11,000-foot-high mesa that remains among the region's most oft-quoted passages.*

It is a sublime panorama. The heart of the inner Plateau Country is spread out before us in a bird's-eye view. It is a maze of cliffs and terraces lined off with stratification, of crumbling buttes, red and white domes, rock platforms gashed with profound cañons, burning plains barren even of sage—all glowing with bright color and flooded with blazing sunlight.

Everything visible tells of ruin and decay. It is the extreme of desolation, the blankest solitude, a superlative desert.

To the northeastward the radius of vision reaches out perhaps a hundred miles, where everything gradually fades into dreamland, where the air boils like a pot, and objects are just what our fancy chooses to make them. Perhaps the most striking part of the picture is in the middle ground, where the great Water Pocket fold turns up the truncated beds of the Trias [*Triassic*] and Jura [*Jurassic*], whose edges face us from a great quadrant of which we occupy the center. Where the strata are cut off in this way upon the slope of a monocline they do not present to the front a common cliff and talus with a straight crest-line, but a row of cusps like a battery of shark's teeth on a large scale. But even in this relation the Jurassic sandstone is peculiar, for it is here of enormous thickness and so massive that it is virtually one homogenous bed, and the great gashes cut across the fold or perpendicular to the face of the outcrop have carved the stratum into colossal crags and domes. By these tokens we can trace the Water Pocket fold from the eastern slopes of Thousand Lake Mountain around a quadrant, whence its course flies off in a tangent far into the south and is lost to view beyond the Colorado. Its total length thus displayed must be about 100 miles. Across this monocline run the drainage channels which head in the amphitheaters along the eastern front of the Aquarius. It is interesting to note how completely independent are these streams of the structural slopes of the country. They rush into a cliff or into a rising slope of the strata as if they were only banks of fog or smoke. It matters not which way the strata dip, the streams have ways of their own. The Fremont River and the creeks which flow down from Thousand Lake Mountain present a very striking relation to the strata. They at first run very obliquely into the fold, and thence by an equally oblique course run out of it again. Nearer to us Temple Creek [*now called Pleasant Creek*] plunges right into the flexure perpendicular to its strike and in the somewhat uncommon relation of a stream running with the dip of the strata. Still nearer, Tantalus Creek runs across the fold in the same general relation but meanders about within it....

Directly east of us, beyond the domes of the flexure, rise the Henry Mountains. They are barely 35 miles distant, and they seem to be near neighbors. Under a clear sky every detail is distinct and no finer view of them is possible. It seems as if a few hours of lively traveling would bring us there, but it is a two days' journey with the best of animals. They are by far the most striking features of the panorama, on account of the strong contrast they present to the scenery about them. Among innumerable flat crest-lines, terminating in walls, they rise up grandly into peaks of Alpine form and grace like a modern cathedral among catacombs—the gothic order of architecture contrasting with the elephantine....

The view to the south and southeast is dismal and suggestive of the terrible. It is almost unique even in the category of plateau scenery.... The rocks are swept bare of soil and show the naked edges of the strata. Nature has here made a geological map of the country and colored it so that we may read and copy it miles away.

# 2

## WILDERNESS LETTER (1960)

### WALLACE STEGNER (1909–1993)

*Wallace Stegner, Utah's great writer of the twentieth century, came to the Capitol Reef country from his home in Salt Lake City as a teenager in the 1920s and forever after wrote warmly of Fruita and the Waterpocket Fold. By the time he sat down one afternoon in 1960 to ponder the meaning of wilderness in a letter to the Department of the Interior, crafting ideas that led to the Wilderness Act, he was the founding director of the writing program at Stanford and well on his way to a Pulitzer Prize for Fiction (for* Angle of Repose, *in 1972). Stegner needed a scene to ground his eloquent musings about the "wilderness idea." He chose the view from Boulder Mountain.*

*In his vision of the essentials of wilderness, Stegner looks past Capitol Reef to the canyons on the west edge of Canyonlands National Park along the Dirty Devil River, an outpost along the Outlaw Trail that came to be called Robbers Roost. Utah's favorite bandits, Butch Cassidy and the Sundance Kid, along with their "Wild Bunch," disappeared into these remote canyons after robberies in the last years of the nineteenth century. Lawmen never penetrated their hideout.*

I grew up on the empty plains of Saskatchewan and Montana and in the mountains of Utah, and I put a very high valuation on what those places gave me. And if I had not been able periodically to renew myself in the mountains and deserts of western America I would be very nearly bughouse. Even when I can't get to the back country, the thought of the colored deserts of southern Utah... is a positive consolation. The idea alone can sustain me. But as the wilderness areas are progressively exploited or "improved," as the jeeps and bulldozers of uranium prospectors scar up the deserts and the roads are cut into the alpine timberlands, and as the remnants of the unspoiled and natural world are progressively eroded, every such loss is a little death in me. In us....

Let me say something on the subject of the kinds of wilderness worth preserving. Most of those areas contemplated are in the national forests and in high mountain country. For all the usual recreational purposes, the alpine and the forest wildernesses are obviously the most important, both as genetic banks and as beauty spots. But for the spiritual renewal, the recognition of identity, the birth of awe, other kinds will serve every bit as well. Perhaps, because they are less friendly to life, more abstractly nonhuman, they will serve even better....

Just as a sample, let me suggest the Robbers' Roost country in Wayne County, Utah, near the Capitol Reef National Monument. In that desert climate the dozer and jeep tracks will not soon melt back into the earth, but the country has a way of making the scars insignificant. It is a lovely and terrible wilderness, such a wilderness as Christ and the prophets went out into; harshly and beautifully colored, broken and worn until its bones are exposed, its great sky without a smudge or taint from Technocracy, and in hidden corners and pockets under its cliffs the sudden poetry of springs. Save a piece of country like that intact, and it does not matter in the slightest that only a few people every year will go into it. That is precisely its value. Roads would be a desecration, crowds would ruin it. But those who haven't the strength or youth to go into it and live can simply sit and look. They can look two hundred miles, clear into Colorado: and looking down over the cliffs and canyons of the San Rafael Swell and the Robbers' Roost they can also look as deeply into themselves as anywhere I know. And if they can't even get to the places on the Aquarius Plateau where the present roads will carry them, they can simply contemplate the idea, take pleasure in the fact that such a timeless and uncontrolled part of earth is still there.

These are some of the things wilderness can do for us. That is the reason we need to put into effect, for its preservation, some other principle than the principles of exploitation or "usefulness" or even recreation. We simply need that wild country available to us, even if we never do more than drive to its edge and look in. For it can be a means of reassuring ourselves of our sanity as creatures, a part of the geography of hope.

# 3

## INTO THE ESCALANTE (2007)

LINDA ELIZABETH PETERSON

*Michigan writer Linda Elizabeth Peterson is one of those pilgrims to southern Utah who finds herself standing on the rim of Boulder Mountain, searching for words. She divides her time between her family's fourth-generation farm and her teaching and writes about the forces that tie us to home, however many "homes" we might have. "Into the Escalante" comes from the 2007 anthology* What Wildness Is This: Women Write about the Southwest.

I had moved to Salt Lake City from central Michigan in 1980 drawn by mountains, city life, ragweed-free air, and every myth about reinvention in the West that a Michigan farm kid could possibly have imagined for herself. But—no surprise—the reinventing business had proven to be hard work. So when my new friend, Dawn, proposed a trip into Southern Utah, I was more than ready.

"So what's there?" I asked.

She rattled off the names of national parks—Arches, Capitol Reef, Canyonlands—and some of those western place names that always quicken my pulse: Deadhorse Point, Grand Gulch, Waterpocket Fold, Wildcat Mesa. I had read my *Desert Solitaire*, written a paper about it, and even had Edward Abbey growl at me at a book signing. But what had I really grasped about Utah's geography? Not much. After all, I'd moved to Salt Lake City without understanding that it was a ski town.

Our trip was more than twenty years ago, but in memory I still stand at

Rainbow over the Henry Mountains and Waterpocket Fold, from Boulder Mountain, 1996.

the top of Boulder Mountain, the pine-scented wind in my face. My eyes sweep across the Waterpocket Fold framed by the Henry Mountains hovering in light clouds far off in the east, and then down into the Escalante River drainage at my feet to the south. Nothing, not the Rockies, not the Tetons, not even the Grand Canyon, has prepared me. I am suddenly at the edge of my known world.

What is this? Rock, all rock, naked and alive. Rolling and undulating, thrusting upward and falling away, twisting and tumbling, it shimmers in a hot, clear June sky. From this height, my gaze can only skim the white and reddish-orange tops of the ridges and domes and the canyons cutting randomly and feverishly through the stone, take in their chaos and mystery, utter wildness scoured and scored into the rock. My Midwesterner's mind, shaped by flatness and the square-mile grid, is undone: safe, comfortable categories release and fragment. My breathing, when it returns, is quick and shallow. My limbs loose, I lean into Dawn. If I say anything, it is "God." If I know anything, it is that I am surrendering to this edge, falling into this land, already filling my hands with it.

The Castle, 1975. I've always felt this photograph captures the essence of Willa Cather's words in *Death Comes for the Archbishop*: "that lightness, that dry aromatic odor ... one could breathe that only on the bright edges of the world."

## PORTFOLIO

# THE BRIGHT EDGE

Photographs by Stephen Trimble

Wingate Sandstone wall, Pleasant Creek, 2009.

Juniper snag, Fivemile Wash, 2012.

Rainbow over the Waterpocket Fold, Panorama Point, 2006.

Cottonwood leaf in mud curls, Spring Canyon, 2008.

Cliff and sky, Meeks Mesa, 2014.

Mancos Shale badlands, North Caineville Mesa, 2005.

Glass Mountain and the Temples of the Moon and Sun, Cathedral Valley, 2016.

The Milky Way arcing over Chimney Rock, 2018.

Fremont River in flood, at Capitol Dome, 2011.

Spring Canyon reflections, 2009.

# PART II

## NATIVE HERITAGE

### Culture & Memory

Anglo settlers called the Waterpocket Fold "Capitol Reef" because they saw this wall of red cliffs as a barricade to travel, akin to an ocean reef. Native peoples saw the Fold as home—not an obstacle but a grid of easy pathways created by permanent streams that slice through these same cliffs.

Archaeologists do their best to understand and analyze artifacts and to classify these Native peoples in recognizable traditions. These scientists place Capitol Reef on the frontier between Great Basin and Southwest cultures. The Fremont people who lived along the Waterpocket Fold in late prehistoric times bump up here against the better-known Ancestral Puebloans, the Anasazi, who mostly lived in canyon country to the south. (Since *Anasazi* means "enemy ancestors" in Navajo, scientists now avoid using the term, in deference to contemporary Pueblo people.)

A heads-up on terms that may be unfamiliar: the Archaic Period in the Southwest follows the Paleoindian Period (the first people in North America) and ends with the adoption of agriculture (in the north) and pottery (in the south). Dates for the Archaic vary, beginning around 8500 BC and ending as late as the first few centuries AD. The Formative Period is the archaeologist's designation for the time of the Fremont and Ancestral Puebloans—the centuries of full-blown agriculture and village life that followed the Archaic.

Writer Rose Houk introduces us to the Fremont, named by archaeologist Noel Morss for the Fremont River at Capitol Reef (the river itself named for explorer John Charles Fremont, who crossed the stream near its headwaters in 1854). Archaeologist Steven Simms probes deeper into the "mystery" and diversity of Fremont people.

To bring to life these long-ago times, writers emphasize people over culture. As David Madsen, former Utah state archaeologist, wrote in *Exploring the Fremont* in 1989:

Fremont Indian petroglyphs, Fremont River canyon, 1975.

> During the fifteen hundred years that the Fremont can be distinguished, they produced an archaeological record as rich, yet as enigmatic, as any in the world. The record of how they lived, reacted and responded to the changing world around them, is a mirror of ourselves: all peoples at all times and in all places. It is a record of human behavior. It is a record that is difficult to interpret and difficult to understand. But if we can clarify who and what the Fremont were, we will better understand how and why we act as we do, and what it means to be human in arid country. By exploring the Fremont, we explore ourselves as a people intrinsically tied to these desert lands.

In the 1990s, National Park Service anthropologist Rosemary Sucec interviewed contemporary Indian people from tribes who used the Waterpocket Fold—the Hopi, Zuni, Southern Paiute, Ute, and Navajo. The stories they tell root each culture in a homeland that includes Capitol Reef National Park. Sucec's work marks a new recognition of inclusivity and collaboration by park managers.

These tribal connections to Capitol Reef country continue. Small communities of Southern Paiute people still live west of the park in Koosharem and Joseph and tell stories of their grandparents' use of the canyons of the Waterpocket Fold.

We don't have to reach back centuries in either Native or non-Native memory to touch the last Ute/Paiute Sun Dance in the Capitol Reef area, held at Fish Lake, source of the Fremont River, in 1929. Rancher Dee Hatch of Bicknell was ten years old that year. Seventy years later, in a 2000 interview with writer and backcountry authority Steve Allen, Hatch remembered witnessing this last Sun Dance, the most important spiritual ceremony (even without the chest-piercing practiced by Plains tribes) in the Ute tradition. The Native people who led the dance at Fish Lake were the same people who claimed Capitol Reef as traditional territory, as sacred ground:

> All the Indians in Utah had an invitation, I think. They had a place out by Bowery Creek, up in the trees in the quaking aspen. They had a clearing up there, and that's the first I'd seen Indians wear braids. We stayed on into the night after dark, and they built this fire and had a nice clearing in the trees and they started dancin' and singin' and all gathered around this campfire and then here come the contestants and they come out of the trees there somewhere and started dancin'. And they had this pole in the middle of the clearing and they'd dance up and back, up and back, and they had feathers on top of the pole and they had a whistle in their mouths, and they were stripped down to the waist, these young men. We noticed with all interest, and in later years, I didn't think much of it. But then I found out that was the last time they ever had a Sun Dance.

In the final piece in "Native Heritage," historian Robert McPherson recounts the odyssey of Capitol Reef's most famous artifacts, the Pectol shields, which found their way home to the Navajo on a winding path that mirrors the journeys of Native people themselves. This repatriation remains the pivotal gesture of respect toward Native culture at Capitol Reef.

Rock art, Fish Creek Cove, 1993. Archaeologist Noel Morss visited these spectacular pictographs and petroglyphs in 1928, and the National Park Service included this Fremont living site near Torrey in their 1935 proposal for a Capitol Reef National Monument. Vandals had already dug the site and defaced the cliff, and so officials chose not to grant monument status to Fish Creek Cove when finalizing the 1937 boundaries.

# 4

## DWELLERS OF THE RAINBOW (1988)

ROSE HOUK

*Flagstaff, Arizona, writer Rose Houk came to the Southwest to work as a park naturalist at Grand Canyon National Park. She has been writing for national park natural history associations for more than thirty years, including two books about Capitol Reef (both excerpted in the reader; also see Chapter 49). Her goal: "to instill curiosity and affection for the natural and cultural world we all share, and thus inspire its protection."*

It is a sultry July day, and as usual the desert is empty. We have followed a creek downstream, crossing and recrossing in detours to examine promising sandstone walls burnished coppery black by desert varnish.

At one point the broad streambed narrows, cutting deep, sinuous swirls into pink rock. Water tumbles playfully through this miniature gorge, inviting us to linger. I walk over to the north wall of the canyon, and as I approach, the sound of flowing water seems to be coming from *within* the rock. An eerie sound. I walk closer, then back away, and the sound changes. I realize that I am hearing the echo of the stream behind me.

And just in front of me, on the face of the canyon wall at eye level, are etched four simple V-shaped figures, their outlines faint from weathering. These figures weren't here by accident; someone else, years ago, also thought this a special place. Whoever had drawn them was marking this place. My imagination soared. Would these have been images of spirits? Water spirits perhaps? In a dry desert such as this, the magical sound of running water would have been worth memorializing.

Bedrock metates, Fremont Culture living site, Pleasant Creek, 2008.

In that canyon of imaginings, we had been treated to one of the finest outdoor galleries in the world. This drainage and other perennial streams in Capitol Reef National Park offer outstanding examples of Indian rock art, much of it executed by people known to archeologists as the Fremont. They lived here about a thousand years ago, then disappeared. Their beginnings, here in this remote corner of the American Southwest, are as uncertain as their end. But they left behind clues about their culture that we can partially decipher and haltingly reconstruct.

The Fremont were named for the Fremont River that flows through the northern section of Capitol Reef, an especially scenic part of the 100-mile-long swell of rock called the Waterpocket Fold. Some consider this area the heartland of the Fremont people, for here they were named and defined as a major prehistoric culture with an identity all its own.

Bits and pieces of the Fremont puzzle have been put in place over the past fifty years, revealing a people who lived intimately with their environment. To survive they had to be highly mobile, but they did settle down long enough each year to plant and harvest crops of corn, beans, and squash. The

Fremont lived in caves or small, unassuming houses, sometimes in villages; they did not build elaborate cliff dwellings or pueblos like their contemporaries to the south. Baskets and pots, tools and weapons, and occasionally clothing are the principal material remains that tell us something of their everyday lives.

Aside from their fantastic rock art, the Fremont are by some standards an "unglamorous" people. Their strong suit, however, was knowing how to live in the valleys, deserts, and mountains of the Capitol Reef area. They had to locate the best spots for farming on the stream bottoms, and know when to journey to the uplands to gather pine nuts. They would have discovered which rockshelters and overhangs stayed dry and warm in winter, and where they might find a dependable spring or a small pothole of residual water during the driest times of the years. By trial and error, they learned hundreds of plants and their uses for food, fiber, and medicine. In the incredibly rugged environment, they had to hone their instincts to outwit prey animals, their only source of meat. That they survived here for more than eight centuries attests to their flexibility, mobility, and ingenuity.

It was up to an easterner, unaccustomed to the desert, to first determine that Capitol Reef held evidence of a distinctive culture that had been largely unknown to archeologists. What he and others have found in the last half century now forms the basis for our knowledge of the Fremont Culture.

## "Discovery"

Noel Morss created a quandary for future archeologists by defining the Fremont Culture. They now talk about "the Fremont problem"—the problem being that the word "Fremont" means different things to different people. The Fremont identity crisis continues to stimulate unending debate among the experts.

When Morss came to Capitol Reef in the summers of 1928 and 1929 he could not have known what his discoveries would later mean. He was part of the Claflin-Emerson expedition, named for Mr. and Mrs. William H. Claflin and Mr. and Mrs. Raymond Emerson, who donated money to Harvard University's Peabody Museum for the expedition.

Using the towns of Escalante, Green River, and Moab as jumping off points, Morss and fellow researchers set out to explore the huge, empty area between the Uinta Mountains on the north and the Colorado River on the south.

They worked in country that is still some of the most remote and inaccessible in the nation. They crossed rivers by ferry because there were no bridges at the time, pulled their horses out of quicksand, scrambled up steep canyon walls, and dug under the searing summer sun. When they could use automobiles, they encountered bad roads that were usually impassable in wet weather. Local residents more familiar with the region and its rigors acted as guides, among them David Rust, known best for his efforts at starting Phantom Ranch at the bottom of the Grand Canyon.

Morss, a lawyer-turned-archeologist, spent most of his time in the Fremont River drainage around Torrey and Fruita, Utah, including the major tributaries of Oak and Pleasant creeks. What he found during the two summers here revealed, in his words, "an unexpected and interesting situation." He had expected to encounter the same kinds of remains as had earlier archeologists working in nearby areas.

Instead, "quite to the contrary, the Fremont drainage proved to be the seat of a distinctive culture." In his landmark monograph published in 1931, *The Ancient Culture of the Fremont River in Utah*, Noel Morss first recognized and defined the Fremont Culture. He described it thus:

> This culture was characterized by cave sites with a slab cist architecture…by a distinctive unpainted black or gray pottery; by the exclusive use of a unique type of moccasin; by a cult of unbaked clay figurines…[and] by abundant pictographs of distinctive types.

Though these people had obviously been influenced by the Anasazi—pueblo builders living to the south, east and west—the Fremont were unique in the characteristics Morss noted.

A great many artifacts had already been removed by the time Morss and other archeologists came on the scene. In 1892 a man named Don Maguire of Ogden, Utah, was commissioned to collect archeological material for Utah's exhibition at the Chicago World's Fair in 1893. Along with a Salt Lake City man, he dug a "quantity of fine objects," including a "mummy," from Capitol Reef, items which did go on display at the exposition, according to early park superintendent Charles Kelly.

Three roaming French archeologists visited the area, scratching their names and the date "1893" on a cliff along the Fremont River. The trio apparently excavated four or five small rooms, but left no other information. As settlers moved into the Torrey area in the late nineteenth and early twentieth centuries, the nonscientific excavating continued. Hide shields, buckskin garments, necklaces, and other choice artifacts could be viewed in "museums" in the homes of local residents.

But it was Noel Morss, and the dauntless Claflin-Emerson Expedition, who finally began to put these finds into a context that had some meaning. As an indirect result of their work, in August 1937 President Franklin D. Roosevelt set aside Capitol Reef National Monument for its "objects of geologic and scientific interest."...

## Dressing up

The Fremont were known by their footwear. They made a moccasin so distinctive that its presence alone says "the Fremont were here."

The moccasin's method of construction makes it unique and original to the Fremont. Three pieces of hide were cut from the forelegs of a deer or antelope. The dewclaws—the claws of vestigial toes—were left on, possibly to function as hobnails. In making the Fremont moccasin, a broad sole was cut in the shape of a foot and two upper pieces were attached to it. These pieces, stitched on with sinew along the sides and in the front of the moccasin, extended and overlapped at the back of the foot. Buckskin or dogbane ties tightened the shoe on the foot.

Noel Morss found such moccasins in Capitol Reef. The "informality in their manufacture" made each pair at first appear to be different. But on closer study, he said, "they turn out to conform fairly closely to one plan." He found no trace of fiber sandals like those commonly worn by contemporaries of the Fremont. Among the moccasins were some with the fur still on the hide, with grass and mud insoles to soften and warm the inside. Morss hypothesized that the use of hide was a response to the cold climate in which the Fremont lived. Winter temperatures in Capitol Reef can indeed make even the toes of the well-booted feel like brittle china.

Moccasins were valued possessions to the Fremont; at least they must have been hard to come by, for many were found patched multiple times.

Before they learned to leave the excavating to archaeologists, Torrey residents Ephraim Pectol and Charles Lee collected Fremont artifacts in the canyons surrounding Capitol Reef in the 1920s. Lee collected this 16½-inch-long cradleboard with its carefully nested miniature Fremont figurine. This remarkable object now resides in the Utah State University Eastern Prehistoric Museum, Price, Utah. Photograph by François Gohier.

Here was a suitable job for a long winter's evening, seated by the fire under a cozy overhang, listening to the old folks tell their timeless stories....

A well-dressed Fremont man or woman was bedecked with ornaments—tubes, pendants, and rings made of bone, teeth, shell, and pottery. The bone pieces are the long bones of small mammals and birds, and have been polished to a sheen. Stone such as jet from the nearby Henry Mountains or turquoise obtained through trade perhaps would have been saved for the finest jewelry. Flicker and meadowlark feathers made handsome headdresses and bands, though only a few have ever been found.

The only fashion clues we have other than these artifacts come from Fremont figurines and rock art. If these in fact indicate everyday dress rather than ceremonial or mythical garb, it appears that women may have worn "aprons" and the men breechcloths. The figurines and the human forms drawn on rock also wear necklaces and sport a variety of head adornments. Indeed, Fremont rock art may tell us a great deal about the people's lives. As one pundit put it, the Fremont, it seems, abided by the dictum "Leave only moccasin prints, take only pictographs."

5

# FREMONT PLACES, FREMONT LIFE, FREMONT PLACE (2008)

STEVEN R. SIMMS

*Archaeologist Steven Simms taught at Utah State University for thirty years. In addition to his scholarly research, he's written popular books about the prehistory of the American Desert West, including* Ancient Peoples of the Great Basin and Colorado Plateau *(from which this excerpt comes) and* Traces of Fremont: Society and Rock Art in Ancient Utah, *with photographer François Gohier. He recently rediscovered a 4.5-mile irrigation system shown to Noel Morss on the slopes of Boulder Mountain in 1928. With modern science, Simms can date these canals, possibly built by the Fremont and used in surprisingly recent times, 1500–1700, suggesting continuity with the historic tribes of the Southwest.*

The Fremont period was a sea change. The ancient heritage of the Archaic forager societies became enmeshed in continental upheaval that brought immigrant farmers and a new way of life to the West. The Fremont was a culture—there were some broad, unifying themes in the rock art, ceramic design, basketry, architecture, and the use of space. But the landscape was large, and there was surely a mosaic of ethnic groups, tribal enclaves, linguistic variation, and shades of difference in lifestyle.

The earliest Fremont landscape, perhaps 200 BC, was a frontier inhabited by Late Archaic foragers and a few explorers from the Southwest. Others followed, and by AD 500 there was a smattering of farming outposts in a wilderness of foragers. After AD 900, the landscape became a sea of farmers. People gathered into villages, hamlets, and farmsteads as they had never done before. In the best watered areas there were true villages of a dozen to several dozen homes. But most Fremont villages were small affairs. Archaeologists call the small Fremont farmsteads "rancherias"; often they consisted only of a single residential structure, but more often there were a few homes clustered together. Among the homes were outdoor activity areas, and probably ramadas for shade. There were racks for hanging things, caches of gear, piles of raw materials, and the refuse dumps that humans everywhere create. Embedded in some of the villages were communal storage structures made of adobe. In other cases storage was kept inside the houses and was hence less public. In still other intriguing instances, storage was remote from the residences and evidently hidden in cliffs for defense.

The primary Fremont dwelling was the pithouse. Basement-like structures, pithouses were capacious, functional, warm in the winter, and cool in the summer. Pithouses were excavated into the ground from a few centimeters to over a meter deep. A log and pole roof supported by vertical posts (a four-square pattern is common) extended the structure well above the ground. When finished with various arrangements of smaller wooden roofing material and finally an earthen seal, pithouses looked like low mounds, or truncated pyramids, with a flat roof that provided a convenient and elevated outdoor living space. Pithouses had central heating, places for storage, and a ventilation system. Shallow, cheap-to-build pithouses had been employed in some parts of the Basin-Plateau for the last 2,000 years, but the extra settling that came with farming brought them into widespread use and escalated the investment made in pithouses during Fremont times. . . .

Despite the number of pithouses often found at Fremont sites, only a few were used at one time. This is true of virtually all Fremont village sites, and we should reject the temptation to see these villages as large just because the count of pithouses runs into the dozens or even hundreds. There were indeed large villages that were home to hundreds of people, but the images conjured by the terms "hamlet" or "rancheria" are a far more accurate description of life for the vast number of Fremont people scattered across the region. . . .

The Fremont were settled, but they were also on the move. A cycle of habitation, abandonment, and relocation structured life and place. This was

Fremont Indian petroglyph of bighorn sheep, Pleasant Creek, 2009.

a repetition that played out over the lives of individuals, as well as a pattern expressed over centuries and hence generations of people....

The lifespan of an individual pithouse was only a few years, and structures that survived more than a decade were rare. People returned and place was a magnet across generations even as particular places went unused for a time....

Fremont rock art reflects the extensiveness of their presence across the region. A distinctive ensemble of anthropomorphic figures, wavy lines, spirals, dots, as well as naturalistic and stylized animals seems to occur in every canyon in the region. The amount of Fremont rock art dwarfs anything from the eons of the Archaic. Fremont rock art exhibits regional styles, but there are common motifs and a unifying manner of execution. Fremont rock art exemplifies the notion that a land once filled with foragers now bristled with enclaves of farmers....

## The desert and the sown

The centuries between AD 800 and 1300 saw a dramatic rise in population size, with the precise timing depending on location. Many now rural areas such as Castle Valley, the Sevier Valley, and Parowan Valley were more widely populated than the same areas are today. The Wasatch Front may well have been the most populous part of Utah, just as it is today. Archaeologists have a dual personality when it comes to the Fremont. On the one hand, we have long known they were farmers. When Noel Morss prospected along the Fremont River near the towns of Loa and Torrey in the 1920s, local farmers showed him the ancient Indian irrigation ditches. Nearly every Utah town founded by Mormon settlers between Brigham City and Cedar City is built on top of Fremont homes and fields, because both peoples chose the best places to farm with the least effort.

On the other hand, archaeologists realized the Fremont were not as married to farming as were the Anasazi of the Southwest. The Fremont were seen as a "northern periphery" relegated to part-time hunting and gathering by the harsh environment of the high plateaus. This became the stereotype of the Fremont....

Perhaps Fremont life was more dynamic than we thought. Is it possible that someone might be born in the desert as a forager, grow up there, and as a young adult find himself marrying into a village and eating maize for the rest of his life? Or perhaps after a few poor years of farming a village might break up, with some families being sent to live with relatives where people relied on a mix of farmed and wild foods. Archaeologists have long wondered about these things, and it began to look as though the scale were tipping toward a dynamic between the desert and the sown....

The DNA and activity patterns show that some of the differences archaeologists saw were reflections of the life history of individuals, rather than categorical differences between people who lived one way or another all their lives. It is the same story of unity as told by the distinctive tradition of Fremont basketry—they were all of a cloth. The Fremont cultural tradition was a tapestry of local ethnicity and variations in ways of life woven together by the social connections that came from the flow of people across the landscape.

Archaeologists see much more than a simple story of foragers settling down to become farmers. The farming life indeed reduced residential mobility as long as people kept farming in a particular place. But excavations of Fremont sites show that most were not occupied long before being abandoned. Absences were sometimes brief, perhaps only a few years, before people returned.... [*And Rosemary Sucec learned in her interviews with the Hopi that they never abandon places. She told me, "Their cosmologies directed them to leave their 'footprints.' They come back to those places on pilgrimage, sing and do ceremonies related to their stopping places. Any ancestors who died there, their spirits still remain."*]

To modern ears, the term "abandonment" sounds like a failure to progress, but perhaps this is the wrong perspective. Abandonment was part of their success. It was part of a sense of place characterized by rhythmic tempo anchored to the land. Fremont farmers flowed over places as the patchwork quilt of arable land changed year to year in response to myriad cues. It might have been the depth of mountain snowpack and consequently the spring and summer runoff. Or it might have been minor differences in elevation, where only 100 meters mattered. A plot might fail or succeed, depending on whether the first frost arrived in August or September. Even the annual exposure of land to sun and shade changed. In some years, a sunny southern exposure might have fried the fields while a nearby northern aspect at the base of a slope leaking moisture might have saved it. In other years it could have been the reverse.

Where did people go when they abandoned places? The first choice was to move to other plots controlled by their family, lineage and alliances among communities. A portfolio of fields in a variety of small settings hedged risk. Over decades, the plots of arable land might move up and down slope as the climate changed. During drought, only the best places might produce, and as population grew, previously unexploited plots could be added. But this approach worked only until the 11th and 12th centuries, when all the arable land was used. The social fabric mediated a dynamic between the desert and the sown, the inhabited and the abandoned, and the tempo of life and place.

6

# FULFILLING DESTINIES, SUSTAINING LIVES (2006)

ROSEMARY SUCEC

*Rosemary Sucec is one of twenty cultural anthropologists who work for the National Park Service. Based at Glen Canyon National Recreation Area, she oversees all cultural resources associated with the canyons along the Colorado River–Lake Powell corridor. Sucec documented American Indian histories and resource uses at Capitol Reef through interviews and archival research to create* Fulfilling Destinies, Sustaining Lives: The Landscape of the Waterpocket Fold, *our most complete source for understanding Native relationships with the park landscape.*

While a study of archeology at the park can relay ecological understandings of how native peoples survived in and adapted to the environment, consultation with associated tribes opens up worlds of meaning that humans through time invest in a landscape.... Too often, Native Americans have been portrayed as merely part of archeology and history—if not otherwise omitted from a presence—and rarely as active players whose heritage preservation is a major responsibility of national park operations.

For example, through consultation with the Pueblo of Zuni, we learned about an epic saga of emergence and millennia-long migrations. The migrations began when their ancestors were hunters and gatherers as early as the Paleoindian and into the Archaic period. One of the Zuni medicine societies headed north potentially into Utah and the Capitol Reef environment. In so doing, ancestors blazed trails and created camping sites that became today's archeological sites. Members of this society used plants,

Fremont Indian petroglyph, Hickman Bridge Trail, 2006.

animals, and minerals to aid them in survival. They scribed images on rocks as maps and for other purposes.

When the Zuni consultants came to Capitol Reef National Park they had never seen it before. Yet, when consultants saw deposits of crystals (though outside the park boundaries) of ceremonial significance to this medicine society, as well as ancient symbols of trails etched on rock, these tangibles assumed a world of meaning. They acted to confirm the historical narratives stored in memories, transmitted orally through hundreds of generations, which suddenly became reinvigorated once they arrived at Capitol Reef. Simultaneously, the National Park Service (as well as the state of Utah) became aware of a potential chapter in its well-aged history, which includes 96 archeological sites of Archaic Period heritage. Any sites dated to this timeframe within the park or on adjacent lands could have been those of their ancestors.

The Hopi Tribe shares a common history with the Pueblo of Zuni. They, too, emerged and began thousands of years of migrations. However, their history with Capitol Reef starts with their ancestors who were farmers, thousands of years later than when Zuni ancestors might have inhabited the landscape. Like the Zuni, contemporary members of Hopi came to Capitol Reef with pre-existing knowledge of the symbol system their ancestors left on rock faces. Though they had never been to Capitol Reef, they immediately recognized these ancient symbols, began reading the rocks, and conveyed to the Park Service yet another saga. The image of the deity who directed them to farm, to conduct rounds of migrations, and to leave (archeological) evidence of their stay was etched at numerous sites within the park. Where this deity is portrayed, farming took place at the site. They saw icons of their clans on rocks that affirmed their pact, their chosen vocation, as well as conveyed the histories of those clans. Other symbols told about local agriculture, provided maps of the region, and formed portions of altars for various ceremonies.

Ancestral Hopi beliefs essentially represented a religious revolution spread by migrant farmers. These beliefs as much constructed Formative sites at Capitol Reef, as did the needs of daily living. They explain why elsewhere, at places such as Grand Staircase–Escalante National Monument, we see a conscious process of reoccupation and remodeling residences. Clans were directed to stay temporarily then continue migrating. The Park Service, in turn, came to have a fuller appreciation of why these farmers came to and stayed in the region or then left. That knowledge, encoded in writing on rocks, was deciphered through consultation with the elders of the Hopi Tribe. The Hopi Tribe relayed to us that it was their ancestors who created the farming sites within the Fold, that is, they had a role in making and using all Anasazi- and Fremont-style sites. 169 Formative Period sites were identified in archeological survey.

The background knowledge gained through interviews with contemporary Paiute and Ute about the Capitol Reef environment, combined with fragments from historical records, reveals not an epic involving spiritual mandates, but humans' relationships with one another, as much as with the environment where they made their living. Collectively the evidence helped us move beyond a one-dimensional view of Numic-speaking ancestors as food seekers. [*These Numic people included in Utah the Shoshone, Goshute, Southern Paiute, and Ute peoples.*] What came into relief was a sense of the camaraderie of families locally and regionally.

We get glimpses of spring celebrations, trading rendezvous, as well as the traffic across the Fold to other enclaves to join in social gatherings or grieve deaths. We learned that while early pioneers' view of the Fold as a barrier was true for them, for Numic-speaking ancestors the Fold contained corridors for travel replete with water. Blending sources of evidence also helped us to see how resilient ancestral Paiute and Ute were in coping with oppressive colonizing forces. A glimmer of the spiritual beliefs was revealed in the reference to the Waterpocket Fold as containing a special rock that was bequeathed by a patron spirit and was a place of renewal for those who knew its location.

For traditional Navajo traveling to the land across the Colorado River that includes Capitol Reef, all of the plants, animals, rocks, and other places are imbued with life forces that protect, yield food, heal, and provide for the well-being of the travelers, their families, and livestock back home. These beliefs about the land to the north affect the actions of individual Navajo when they cross the Colorado River and as they continue to travel through the landscape until they return home. Certain etiquette is required, which means the creation of certain types of archeological sites.

Navajo elders also relayed over 300 years of history with the study area when all we had heard was anecdotal stories of horsemen traveling through the park when the settlers lived there in the late nineteenth century. Oral accounts pertaining to the Capitol Reef shields in tandem with radiocarbon dating of the shields suggest Navajo ancestors were here as early as circa 1500 or the sixteenth century. Oral tradition alone hints at an even earlier occupation, perhaps as early as the fifteenth century. This traditional knowledge, new to Capitol Reef (if not the state of Utah), represents opportunities for archeological investigation. Historical events such as the Puebloan Revolt in the 1600s, the Spanish Entrada into Utah shortly after, and the Kit Carson roundup at mid-nineteenth century may have resulted in the Capitol Reef region becoming a place of refuge....

Tribal histories with this environment literally span at least 10,000 years and are continuous into the early twentieth century. Park Service administration, by contrast, encompasses approximately 70 years. And while Indians do not interact with the land the way they did in the past, what remains are the tangible resources, places, and landscapes that are linked with their cultures, including their histories, and that have been documented in and paid homage to with this report. What also remains, but can be inadvertently overlooked, are [*in the words of anthropologists T. J. Ferguson and Roger Anyon, who work with Native nations on cultural heritage preservation*] the multiple tribal "cultural processes of memory and history to renew the links with [these tangible] places [that have been] forgotten, irregularly visited, or occupied by other groups."

In the schema of human association with the Fold, the National Park Service has become the most recent occupant in a long and distinguished lineage of homesteaders. When juxtaposed with historical, legal, and environmental construction of the Waterpocket Fold as a vacant wilderness and as a barrier, this represents a daunting perspective—as well as a significant, yet challenging, management responsibility.

# 7

## SEEING IS BELIEVING

### The Pectol Shields (2012)

ROBERT MCPHERSON

*Historian Bob McPherson taught for forty years at the Utah State University campus in Blanding. His twenty books grow from dozens of interviews with San Juan County old-timers, Indian and non-Indian. This narrative of the remarkable leather shields unearthed in Wayne County in 1926 comes from his* Dinéjí Na`nitin: Navajo Traditional Teachings and History. *McPherson also wrote a biography of John Holiday,* A Navajo Legacy: The Life and Teachings of John Holiday.

*The shields remain controversial. Despite the story told in this excerpt, some experts still believe the shields were made by Ute people, others by the Fremont. Still others believe we can't definitively know their origins.*

Sparks from the piñon and juniper fire rose into the black night sky. Shadows danced on the low alcove's walls, flames flickering with wind currents. Nine figures crowded beneath or stood outside a low overhanging ledge, as some bent forward digging and peering into a hole in the sandy-bottomed cave. There was nothing to distinguish this particular site, a mere four feet by six feet, from any other of the countless crevices and rock niches surrounding the little town of Torrey and what would later become Capitol Reef National Park, Utah. Supervising the excavation was Ephraim Portman Pectol, a Latter-day Saint (LDS) bishop, entrepreneur, and promoter of Wayne

County. His wife, Dorothy; three daughters; their son-in-law, Claude Holt; and three other men assisted with what everyone anticipated to be a Native American burial of some type. Earlier that day Ephraim and Dorothy had discovered a cedar bark covering eighteen inches beneath the sandy floor of the cave. They decided to let family members share the thrill of discovery, returning with them and others in the evening for an enjoyable outing. Growing anticipation accompanied the unveiling, with Ephraim hoping to add something significant to his burgeoning collection of Indian artifacts on display at home.

It was August 16, 1926, twenty years after the US Congress passed the Antiquities Act to protect archaeological sites from collectors and vandals. In southern Utah, however, professional archaeologists as well as avocational pot hunters burrowed into ruins, burials, and any other site that might hold objects left behind by prehistoric Indians. Today, many of the efforts of even the "trained, professional" archaeologists of the time would be classified more as looting than scientific excavation. Collecting was everyone's intent. While large Ancestral Puebloan (Anaasázi) ruins like Mesa Verde (Colorado) and Chaco Canyon (New Mexico) had been under the spade and trowel of the Wetherill family at the turn of the twentieth century, what seemed to be endless smaller sites drew less attention and were easily accessible to local people. In south-central Utah where Pectol lived, the highly developed Anaasázi material culture gave way to the less dramatic Fremont remains. Still, there were objects to be had and no telling what might be unearthed during a dig of discovery.

Scraping away more dirt and removing a four-inch cedar bark covering, Ephraim exposed a circular piece of hide approximately thirty-six inches in diameter. Expecting to find a body with a few primitive tools, his eyes must have bulged when he beheld three buffalo-hide shields painted in dazzling multicolored geometric patterns. Lifting the objects out of the ground and into the flickering firelight, he "unearthed three of the most wonderful shields ever seen by man. As we raised the front shield the design on two shields came to view. For the space of what seemed two or three minutes, no one seemed to breathe; we were so astonished. We felt we were in the presence of the one who had buried the shields. And these words came to me while in this condition: 'Nephites and Gadianton Robbers.'" Beneath the last object was a cone of earth that maintained the convex shape of the tanned leather shields with their arm and neck carrying straps; next was a bottom layer of cedar bark to guard against moisture.

## Pectol's and other LDS interpretations

What Pectol really discovered that evening was the beginning of a controversy that remains to this day. Ephraim Pectol, very much a man of his time, filtered what he saw through what he believed. His initial response to what he saw as Nephites and Gadianton Robbers [*in the ancient Americas, a group of settlers and a criminal gang, respectively, in Joseph Smith's* Book of Mormon] was totally in keeping with his experience as an LDS bishop for sixteen years, steeped in the teachings of the *Book of Mormon*. Three transoceanic crossings of Israelites before the time of Christ, the rise and fall of

Ephraim Portman Pectol and his two daughters, Devona (*left*) and Golda (*right*), ca. 1928–1930, with the three shields discovered in a Waterpocket Fold alcove near Torrey by Ephraim and his wife, Dorothy, in 1926. Photograph used with permission of the Pectol Family Organization.

the Nephite and Lamanite civilizations, the belief that their descendants were directly connected with today's Native Americans, and all of the religious teachings recorded before the fall of their society furnished dramatic fare for interpreting archaeological remains. Mormons living in southern Utah did not hesitate to connect ancient Indian artifacts and sites with these events. Even the discovery of the shields took on religious tones. Dorothy, guided by the Spirit, had directed her husband, who had not received as strong an impression of where to dig: "You must dig into this and you will find something." And he did....

## The shields: Fame, defame, and reinterpretation

A continual proponent of Wayne County, Pectol spread information about the Capitol Reef area's prehistoric treasures. He showed the Noel Morss archaeological expedition a few Fremont sites around Torrey in 1928 and 1929. Morss was the first archaeologist to designate the Fremont Indians as a distinct culture. Pectol and Morss revisited the rock shelter where the shields had been discovered in hopes of finding more objects to provide contextual clues to the shields, but they found nothing. Morss recognized that the shields were anomalous in the archaeological record, and in his report on the Fremont River drainage he expressed "the opinion that these remarkable shields date from comparatively recent if not historic times. This conclusion is based on their uniqueness among objects of ancient origin, on their resemblance to modern Athabascan shields.... The shields, while modern from the point of view of the Fremont culture, may still be old from a historical standpoint."

In addition to archaeologists, other people visited Torrey to see the shields, which Pectol shellacked for protection, as he did many of his other artifacts.... The holdings in the museum drew others to Wayne Wonderland, a new name for the area now known as Capitol Reef and environs. Charles Kelly—writer, researcher, and adventurer—initially came to Torrey to visit Pectol's museum. In 1943 he became the first custodian of Capitol Reef National Park [*then, still a national monument*]. The artifacts were also highly revered among Pectol's family members. His grandson, Neal Busk, remembers playing a Fremont flute in the room above his general merchandise store, something that would surely make an archaeologist shudder.

When asked where he found the shields, Pectol had Joe Covington, his grandson, lead the inquirers to the rock shelter by a circuitous route coming and going, making it difficult to relocate. Covington, with clear conscience, took the people to the spot designated by Pectol, who had intentionally misinformed him. The discoverer regarded the real site sensitive and sacred enough to keep it hidden from all except his closest kin. Covington was surprised years later when a family member led him to the real spot....

[*By the late 1930s, Pectol had loaned the shields to the Temple Square Museum in Salt Lake City. Two shields came back to Wayne County in the 1950s, and finally, in 1965, the Park Service chose to display all three shields at the new Capitol Reef Visitor Center.*]

For forty-six years, the shields rested comfortably in their display case at Capitol Reef, receiving an occasional mention in archaeological journals. Then in 1998, enter for the first time the Native American view. Representatives from Zuni Pueblo visited Capitol Reef and saw some small hide-wrapped bundles containing bone, which they considered sensitive grave objects, on display. At their request the items were removed. While the relatively new Native American Graves Protection and Repatriation Act, passed in 1990, was not cited for removal of the items, the spirit of the law was evident. Under this law the federal government recognized the rights of Indian people to their ancestors' cultural items, such as funerary, sacred, and other cultural objects as well as human remains. Every museum and federal agency was required to inventory and notify Native American tribes of any objects that met these criteria. The artifacts were to be repatriated "expeditiously" to a requesting tribe after it demonstrated cultural affiliation with the artifact. The ruling criteria for this determination were to provide a "preponderance of the evidence based upon geographical, kinship, biological, archaeological, anthropological, linguistic, folkloric, oral tradition, historical, or relevant information or expert opinion...."

In 1998 the federal government required all national parks to remove NAGPRA items from display. The two remaining shields [*on display in the Capitol Reef Visitor Center*] joined the third shield in the National Park Service's Western Archeological and Conservation Center in Tucson. They then became subject to repatriation. The Navajo Nation was the first to submit a claim. By 2002, the Utes in Uintah, the Paiute Tribe of Utah, and the Kaibab

Band of Paiutes had also entered a joint repatriation request, as had the Southern Ute and Ute Mountain Ute Tribes in Colorado. The primary argument offered by all claimants, except the Navajo, was rooted in cultural affiliation. The lands where the shields surfaced were traditionally those of the Paiute and Ute people. The Navajos provided a far more detailed explanation of the origin and use of the shields.

In an effort to maintain impartiality while assigning ownership, the government hired four scholars... to prepare independent studies of the shields. Based on physical evidence that included materials used in construction and decoration, prehistoric rock art, historic tribal locations, cultural practices, and other considerations, each person was to suggest which tribe he or she thought had made the shields or state that affiliation could not be reasonably established. There was no definitive agreement....

## John Holiday and the Navajo interpretation

The Navajo Tribe asked John Holiday, a Navajo medicine man and prime witness, and three other tribal members to testify on behalf of its claim. Born in Monument Valley, Utah, around 1919, John grew up in the traditional environment that eventually led to his becoming a Blessing Way singer and a repository of cultural and historical knowledge. On March 8, 2001, the Navajo contingent met with National Park Service officials at the Western Archeological and Conservation Center. Lee Kreutzer, cultural resources program manager of Capitol Reef and lead investigator for the repatriation claims, held a second meeting with John and two men from the Navajo Nation Historic Preservation Department (NNHPD) in Monument Valley on May 7, 2002. He not only named the individuals who made the shields and told how they got to the burial site, but he also interpreted the meaning of the designs and the powers they held. What follows is a brief summary.

John has knowledge of family members who lived in the vicinity of the No Name (Henry) Mountains, White Face (Boulder) Mountain, and the area southeast of Richfield along the Fremont River. This was prior to the 1860s and the period known to the Navajo as the Fearing Time and the Long Walk, when many were forced into exile at Fort Sumner, New Mexico. This region is not traditionally viewed as Navajo land but rather as Ute territory. The Navajos, who were following a nomadic, herding, hunting-and-gathering

John Holiday, Navajo medicine man, 2013.
Photograph by Gavin Noyes, Utah Diné Bikéyah.

lifestyle at this time, were constantly searching for verdant lands. John's grandmother, Woman with Four Horns (named after the type of goats she raised), herded sheep in this area and was able to name other Navajo families living close by.

According to John, the powers of the shields were created spiritually at the beginning of the earth, as were medicine bundles used for healing and protection. The holy people are the ones who control their powers and assisted the first person who made the physical shields in question. These objects are both a representation of nature's invisible powers and a living entity that can control and use those powers on the Navajos' behalf. The shields had been in the possession of eight generations of medicine men before the Fort Sumner experience. Objects of this nature are viewed as alive and so must be "fed" or renewed with songs, prayers, and pollen or sacred stone (ntł'iz) offerings. Transmission of the shields, as with medicine bundles (jish), is made from one medicine man to another, not within a single family.

Prior to the time of the Long Walk, a series of medicine men—some of whom were from the Capitol Reef area—held the responsibility to renew and safeguard the protective shields. Many Goats with White Hair created the shields, making them in the Kaibab Mountains in a thick pine forest with a circular clearing. Custody of the shields went to Man Who Keeps His Mouth Open, then Yellow Forehead, Tall Skinny Man, Man Who Wants to Sit Down, Side Person, Man Who Plays with the Wooden Cards, Man with Metal Teeth, Ropey, and finally to Little Bitter Water Man. The shields had different names: Earth Protective Shield, Heaven's Protective Shield, Mountain Protective Shield, and Water's Protective Shield. They were decorated with the likeness of the invisible protective powers held by the natural entity named. For example, the Earth, a living being, has its own shield of protection; by copying its elements, the Navajo can likewise draw upon its protective powers....

The powers of the shields, once renewed, could not be penetrated by bullets and arrows or evil and witchcraft. They provided protection against all things that can harm a person or a group under its power.

When the U.S. military and its Indian and New Mexican auxiliaries warred against the Navajos to the south, those living near the Henry Mountains remained safe and escaped exile to Fort Sumner: "The Navajo were put in the 'heart' of the shields and were safe" [*John Holiday told Lee Kreutzer*]. "They were not captured. They remained hidden in the Henry Mountains and surrounding area where these sacred shields were and so were never caught.... They did not go to Fort Sumner because they lived closer to the sacred shields. It is said that these shields were often taken to other parts of our land, throughout the Navajo communities, just as the sacred mountain soil medicine bundle [jish] is carried around."

During this time, however, as the people evaded detection, Ropey and Little Bitter Water Man had control of the shields and wanted to prevent their capture. Little Bitter Water Man hid them and left the area. He became sick and died without telling anyone where the shields were hidden, causing them to be "misplaced." The powers were neglected, their influence waned, and the invasion of Navajo lands and capture of the people resulted in the four-year imprisonment of over 8,000 Navajos. They had lost their protection. With the rediscovery of the shields, an opportunity to renew these powers became possible.

## Repatriation and aftermath

Based on an evaluation of all claims submitted by various bands and tribes, Kreutzer recommended that the shields be returned to the Navajo Nation. Her decision took immediate fire but was based in the very heart of the repatriation process and Native American thought, in this case Navajo....

As for the claims of the various tribes, many of them were "non-specific and not linked to a particular oral tradition that they could share with the National Park Service," whereas those of the Navajo Nation were. The others felt the place where the shields had been found was historically their territory and that they had a responsibility to a higher power to reclaim what they believed was theirs. Even though they shared dependence on oral history and mnemonic devices—"external visual symbols such as costumes, masks, totems, design motifs, petroglyphs, and pictographs as memory aids"—as did the Navajo, most of this "information management strategy" was lacking in their claim. Thus Kreutzer awarded possession of the shields to the Navajo, who best fit the criteria outlined in NAGPRA.

Capitol Reef superintendent Albert Hendricks approved the transfer on August 1, 2003. John Holiday and Marklyn Chee of the NNHPO [*Navajo Nation Historic Preservation Office*] drove to Tucson, retrieved the shields, and brought them to Window Rock on August 7. Chee felt it was an emotional journey. He drove while Holiday sang and prayed over the shields: "The songs were to revive them and tell them 'You're home.'... It felt like a good thing to bring them back." Because they are sacred objects, they are not on display and continue to be subject to controversy.... Chee... reiterated that the shields "are kept in the museum and are not on display. I repeat, not on display.... They are for ceremonial purposes and they have been reintroduced into ceremonial use...."

The odyssey of the Pectol shields began that August night eighty-six years ago and has continued ever since. Initially viewed as proof of the *Book of Mormon*, the shields have seemed out of place in the Capitol Reef area. The shields... have been used to prove the validity of LDS temple ceremonies, Native American ceremonies, and Masonic rituals. Archaeologists have used the shields to assert a Northern Plains origin for the Fremont culture, to refute such claims, and to suggest trade connections between southern Utah and the Plains. They have been viewed as the reason the Navajos were defeated in the 1860s and spent four agonizing years at Fort Sumner. Until 2003, the shields were the subject of a Native American repatriation battle.

Underlying the controversy is a fundamental issue. Varying perceptions have applied more "color" than is encountered on the physical objects. Even those who pride themselves on their interpretation of factual evidence struggle with finding a definitive answer as to creation and ownership. In the meantime, Navajo medicine men occasionally remove the shields from their containers to renew their powers, feeding them with prayers and songs. Harmony and protection result. Others, not of the same inclination, want an opportunity to study the shields further. Conceivably, this could open the door to lawyers, standing in the wings and ready to apply their perception of who holds proprietary rights to these objects. The odyssey could continue. The crux of this controversy hits at the heart of how we write history, "do" archaeology, and honor religious practices—for "seeing is believing."

Halls Creek and Lake Powell, from Hall Mesa, 2011.

# PART III

## EXPLORATION

### Walking on the Flinty Mountains

The first non-Native explorers who encountered Capitol Reef didn't know exactly where they were. John Charles Fremont's unlucky Fifth Expedition ground to a halt after a snowy passage through Cathedral Valley and over Thousand Lake Mountain. The expedition photographer, Solomon Carvalho, nearly froze to death on this 1854 journey, and his narrative of the men's experience is riveting. Carvalho's daguerreotype of "Natural Obelisks" in the park's Cathedral Valley became the first photograph taken in Capitol Reef. The modern photographer Robert Shlaer retraced Fremont's route, making his own twentieth-century daguerreotypes of the same scenes photographed by Carvalho.

After Fremont came John Wesley Powell's field crews, much better at turning exploration into science. Jack Hillers (who became Powell's chief photographer) documents in this journal excerpt the first scientific traverse of the Waterpocket Fold in 1872, when Powell's brother-in-law A. H. Thompson (known as "Prof") took his men down Pleasant Creek. The Fold wasn't their primary destination. The crew was headed for the mouth of the Dirty Devil River, east of the Henry Mountains, where they needed to reclaim a boat they had cached on their Colorado River expedition.

Grove Karl Gilbert came next, delegated by Powell to investigate the intriguing geology of the Henry Mountains. The Henrys turn up in every geology textbook as the place where Gilbert untangled the nature of laccolithic mountains—where igneous rock has intruded between sedimentary beds, pushing the overlying strata into mountainous domes, almost-volcanoes.

Charles B. Hunt, who ran his own pack-train fieldwork in the Henrys as a US Geological Survey geologist in the 1930s, looks back here at the notebooks Gilbert kept in the 1870s and finds insight and excitement in them. In a 1985 interview at his home in Salt Lake City, Hunt told me:

Grove Karl Gilbert, ca. 1910. Department of the Interior/US Geological Survey.

Gilbert is my greatest teacher. Though I never met the man. He had died before I became a cub geologist. But I had his field notebook when I was working in the Henry Mountains.

Gilbert had the ability to argue with himself on paper. He'll put down an idea, and two or three pages later he'll discuss that idea. Pros and cons. On paper. And I also discovered that there were whole sections in his Henry Mountain monograph that are verbatim out of his field notes.

All I had in the way of a topographic map was a copy of the 1875 map made by the Powell survey. We went into the Henry Mountains with a white sheet of paper.

The Plateau taught me astronomy. It's glorious day or night. We could see sixth magnitude stars down there in the Thirties. It was a young man's country. I couldn't take today some of what I took then, but I don't feel regretful because some of the things I can do today I couldn't do then.

Hunt also told me a story about the conflicting names for the river flowing through the park.

> Gilbert refers to the river as the Dirty Devil clear up into Rabbit Valley [*the upper Fremont valley around Loa*]. The maps were all calling it Fremont when we were there, and I wanted to restore Powell's usage, but I knew the Board of Geographic Names would consult the local people because their rule is—local usage prevails. Everybody I knew called the whole river the Dirty Devil.
>
> I knew they would contact Martha MacDougal, the Hanksville postmistress, so I went into her office one day and asked her the name of the river in this canyon below town.
>
> She said, "Oh you mean the Dirty Devil."
>
> I knew I was in.
>
> She said, "Nice people call it the Fremont. But there haven't been any nice people in Hanksville since a long time ago."

Nevertheless, upstream from Hanksville it's still "the Fremont."

Exploration gradually eased into tourism. When experts in local geography and lore like Dave Rust guided wealthy customers like George Fraser (a New York City lawyer fascinated by John Wesley Powell's explorations on the Plateau), the party was out for adventure. Still, they traveled in such unknown country that their early twentieth-century records were often the first we know from these places. We dip into George Fraser's journal of a trip from Cathedral Valley to Caineville.

When LDS apostle John Widtsoe traveled along the Waterpocket Fold all the way to the Colorado River in 1922, he and his party were among the first non-Native travelers to do so. They knew where they were going, but they were still reconnoitering—in this case, headed for Glen Canyon to look for dam sites.

The challenge of Capitol Reef in winter: Chimney Rock, 2018.

# 8

## INCIDENTS OF TRAVEL AND ADVENTURES IN THE FAR WEST (1858)

### SOLOMON NUNES CARVALHO (1815–1897)

*Solomon Carvalho was perfectly content minding his Philadelphia business—painting portraits and making daguerreotypes, using the first widely available photographic process to fix images on copper plates—when John C. Fremont came calling in August 1853. The controversial explorer was headed west for his last expedition, a second attempt to follow the potential railroad route along the 38th parallel in winter. Fremont needed a chronicler, and he hired Carvalho, a Sephardic Jew who was no outdoorsman but who was a well-known photographer.*

*The trip was grueling and got worse after the party crossed the Green River and headed for what would become Capitol Reef National Park. Oliver Fuller, the assistant topographer mentioned in Carvalho's piece, died a few days after the men climbed over Thousand Lake Mountain from Cathedral Valley in the storm recounted here.*

*Fremont and Carvalho and the rest of the expedition managed to push through the snows to reach a meadow where the present town of Fremont sits just west of the park at the foot of Thousand Lake Mountain. Here Colonel Fremont came to grips with reality and decided to cache nearly all his gear, freeing the pack mules of their burdens so his remaining men could ride and survive the rest of the journey westward to the Utah settlements. Fremont carved his name on a cottonwood tree at the cache site, an inscription that survived for decades and gave the town of Fremont its name.*

*Carvalho left the expedition in Parowan, but Fremont continued all the way to California. Carvalho went home to the East Coast and never returned west. His*

Solomon Nunes Carvalho, ca. 1850. Library of Congress.

*account of his travels, the only description of Fremont's final expedition, became a bestseller. Most of Carvalho's photographs were later lost in a warehouse fire, but engravings made for a proposed book by Fremont preserved many of his scenes.*

*Though his expedition failed to convince America to build a transcontinental railroad along his route, Fremont's name remains on prominent display in Wayne County. Fremont's life and career continued for decades, careening from success to failure until he died, nearly destitute, in 1890.*

Mr. Egloffstein, Mr. Fuller, and myself [*the topographer, his assistant, and the expedition daguerreotypist, respectively*] were generally at the end of the train, our scientific duties requiring us to stop frequently on the road. Mr. Fuller had been on foot several days before any of the rest of the party, his horse having been the first to give out. On this occasion, we started out of camp together. We were all suffering from the privations we had endured, and, of the three, I was considered the worst off. One of my feet became sore, from walking on the flinty mountains with thin moccasins, and I was very lame in consequence. Mr. Fuller's feet were nearly wholly exposed. The last pair of moccasins I had, I gave him a week before; now his toes were out, and he walked with great difficulty over the snow. He never complained when we started in the morning, and I was surprised when he told me he had "given out."

"Nonsense, man," I said; "let us rest awhile, and we will gather fresh strength." We did so, and at every ten steps he had to stop, until he told us that he could go no further.

Mr. Fuller was the strongest and largest man in camp when we left Westport [*the Kansas town where the expedition began in the fall of 1853*], and appeared much better able to bear the hardships of the journey than any man in it. I was the weakest, and thought ten days before that I would have given out, yet I live to write this history of his sufferings and death, and to pay this tribute to his memory.

The main body of the camp had preceded us, and they were at least four miles ahead. Both Mr. Egloffstein and myself offered our personal assistance; Mr. Fuller leaned upon us, but could not drag one foot after the other—his legs suddenly becoming paralyzed. When we realized his condition, we determined to remain with him; to this he decidedly objected—"Go on to camp," said he, "and if possible, send me assistance. You can do me no good by remaining, for if you do not reach camp before night, we shall all freeze to death."

He luckily had strapped to his back his blue blankets, which we carefully wrapped around him. In vain we hunted for an old bush or something with which to make a fire—nothing but one vast wilderness of snow was visible. Bidding him an affectionate farewell, and promising to return, we told him not to move off the trail, and to keep awake if possible.

Limping forward, Egloffstein and myself resumed our travel; the sun had passed the meridian, and dark clouds overhung us. The night advanced apace, and with it an increase of cold. We stopped often on the road, and with difficulty ascended a high hill, over which the trail led; from its summit I hoped to see our camp-fires; my vision was strained to the utmost,

Early snow, Cathedral Valley, 1975.

but no friendly smoke greeted my longing eyes. The trail lost itself in the dim distance, and a long and weary travel was before us. Nothing daunted, and inspired by the hope of being able to render succor to our friend, we descended the mountain and followed the trail.

It now commenced to snow. We travelled in this manner ten long hours, until we came upon the camp.

Mr. Egloffstein and self both informed Col. Fremont of the circumstance, and we were told that it was impossible to send for Mr. Fuller.

Overcome with sorrow and disappointment, I fell weeping to the ground. In my zeal and anxiety to give assistance to my friend, I never for a moment thought in what manner it was to be rendered. I had forgotten that our few remaining animals were absolutely necessary to carry the baggage and scientific apparatus of the expedition, and that, with a furiously driving snow-storm, it was almost folly to attempt to find the trail.

While we were speaking at our scanty fire of the unfortunate fate of our comrade, Col. Fremont came out of his lodge, and gave orders that the two best animals in camp should be prepared, together with some cooked horse-meat. He sent them with Frank Dixon, a Mexican, back on the trail, to find Mr. Fuller. We supposed him to have been at least five miles from camp.

There was not a dry eye in camp that whole night. We sat up anxiously awaiting the appearance of Mr. Fuller. Col. Fremont frequently inquired of the guard if Mr. Fuller had come in.

Day dawned, and cold and cheerless was the prospect. There being no signs of our friend, Col. Fremont remarked that it was just what he expected.

Col. Fremont had allowed his humanity to overcome his better judgment.

At daylight, Col. Fremont sent out three Delawares to find the missing men; about ten o'clock one of them returned with Frank Dixon, and the mules; Frank had lost the trail, he became bewildered in the storm, and sank down in the snow, holding on to the mules. He was badly frozen, and became weaker every day until he got to the settlements. Towards night, the two Delawares supporting Mr. Fuller, were seen approaching; he was found by the Delawares awake, but almost senseless from cold and starvation; he was hailed with joy by our whole camp. Col. Fremont as well as the rest of us, rendered him all the assistance in our power; I poured out the last drop of my alcohol, which I mixed with a little water, and administered it to him. His feet were frozen black to his ankles; if he had lived to reach the settlements, it is probable he would have had to suffer amputation of both feet.

Situated as we were, in the midst of mountains of snow, enervated by starvation and disease, without animals to carry us, and a long uncertain distance to travel over an unexplored country; could any blame be attached to a commander of an expedition, if he were to refuse to send back for a disabled man? I say, no, none whatever. Twenty-seven of our animals had been killed for food, and the rest were much reduced, and without provender of any kind in view. If this event had occurred six days later, there would have been no animal strong enough to carry Mr. Fuller into camp.

But suppose he had been disabled while in camp, and unable to proceed, could blame attach to his comrades if he were deserted, and left to die alone? This frightful situation was nearly realized on several occasions. I again answer, no, not any—the safety of the whole party demanded their immediate extrication from the dangers which surrounded them; every hour, every minute, in these mountains of snow, but increased their perils; on foot, with almost inaccessible rugged mountains of snow to overcome, with no prospects of food except what our remaining animals might afford—to stop, or remain an indefinite time with a disabled comrade, was certain death to the whole party, without benefitting him; his companions being so weak, that they could not carry him along....

After we crossed the Green River, the whole party were on foot. The continued absence of nutritious food made us weaker every day. One of my feet was badly frozen, and I walked with much pain and great difficulty; on this occasion my lameness increased to such a degree, that I was the last man on the trail, and my energy and firmness almost deserted me. Alone, disabled, with no possibility of assistance from mortal man, I felt that my last hour had come; I was at the top of a mountain of snow, with not a tree to be seen for miles. Night approached, and I looked in vain in the direction our party had proceeded, for smoke or some indication that our camp was near. Naught but a desert waste of eternal snow met my anxious gaze—faint and almost exhausted, I sat down on the snowbank, my feet resting in the footsteps of those who had gone before me. I removed from my pocket the miniatures of my wife and children, to take a last look at them. Their dear smiling faces awakened fresh energy, I had still something to live for, my death would bring heavy sorrow and grief to those who looked to me alone for support; I determined to try and get to camp, I dared not rest my fatigued body, for to rest was to sleep, and sleep was that eternal repose which wakes only in another world. Offering up a silent prayer, I prepared to proceed. I examined my gun and pistols, so as to be prepared if attacked by wolves or Indians, and resumed my lonely and desolate journey. As the night came on, the cold increased; and a fearful snow storm blew directly in my face, almost blinding me. Bracing myself as firmly as I could against the blast, I followed the deep trail in the snow, and came into camp about ten o'clock at night. It requires a personal experience to appreciate the intense mental suffering which I endured that night; it is deeply engraven with bitter anguish on my heart, and not even time can obliterate it.

Col. Fremont was at the camp fire awaiting my arrival. He said he knew I was badly off, but felt certain I would come in, although he did not expect me for an hour.

My haggard appearance sufficiently indicated what I suffered. As I stood by the fire warming my frozen limbs, Col. Fremont put out his hand and touched my breast, giving me a slight push; I immediately threw back my foot to keep myself from falling. Col Fremont laughed at me and remarked that I had not "half given out," any man who could act as I did on the occasion, was good for many more miles of travel. He went into his tent, and after my supper of horse soup, he sent for me, and then told me why he played this little joke on me; it was to prevent my telling my sufferings to the men; he saw I had a great deal to say, and that no good would result from my communicating it. He reviewed our situation, and the enervated condition of the men, our future prospects of getting into settlements and the necessity there was for mutual encouragement, instead of vain regrets, and despondency; the difficulties were to be met, and it depended on ourselves, whether we should return to our families, or perish on the mountains; he bade me good night, telling me that in the morning he would endeavor to make some arrangements to mount the men.

The next day, he called the men together and told them that he had determined to "cache" all the superfluous baggage of the camp, and mount the men on the baggage animals, as a last resource. Nothing was to be retained but the actual clothing necessary to protect us from the inclemency of the weather.

A place was prepared in the snow, our large buffalo lodge laid out, and all the pack saddles, bales of cloth and blankets, the travelling bags, and extra clothes of the men, my daguerreotype boxes, containing besides, several valuable scientific instruments, and everything that could possibly be spared, together with the surplus gunpowder and lead, were placed in it, and carefully covered up with snow, and then quantities of brush to protect it from the Indians. I previously took out six sperm candles from my boxes, and gave them to Lee, the Colonel's servant, in charge; they were subsequently found most useful. A main station was made at this place, so as to be able to find it if occasion demanded that we should send for them.

The men now were all mounted; a large mule was allotted to me, and we again started, rejoicing in having animals to carry us. After this, every horse or mule that gave out, placed a man on foot without the possibility of procuring others, and it was necessary in consequence of the absence of grass to allow the mules to travel as light as possible; we therefore relieved them frequently by walking as much as we were able.

# 9

# SIGHTS ONCE SEEN

## Daguerreotyping Fremont's Last Expedition through the Rockies (2000)

ROBERT SHLAER

*Robert Shlaer taught himself the nearly lost art of making daguerreotypes in the mid-1970s and gave up his academic position as a neurophysiologist at Northwestern to return to the West, where he'd grown up in New Mexico. He perfected his craft, outfitted his minivan as a mobile laboratory, and set out to re-create the path and the photographs of Solomon Carvalho. After five years of work, Shlaer published* Sights Once Seen, *recounting the two photographic journeys 140 years apart, including this section on Carvalho's crucial time in Capitol Reef.*

After passing Wild Horse Butte, the expedition was finally able to turn west, find more water by crossing Muddy Creek, and skirt the Moroni Slopes to ascend by Salt Wash into Cathedral Valley. On the way they would have seen the massive Factory Butte in front of the Henry Mountains to the south. John Wesley Powell and his men are usually given credit for discovering the "Unknown Mountains" in 1872, but Frémont and his topographer Egloffstein had them in sight for weeks in early 1854. The range is placed correctly on Frémont's map, labeled "Réjas del oso" [*"Bear grills or bars"*]. Unfortunately, Frémont did not publish this observation in a timely manner, leaving it for Powell to garner the honor of discovery.

Following Middle Desert Wash while ascending Cathedral Valley, Frémont and his weakening band could not have missed the fabulous sandstone cliffs and formations which give this region its visual interest, the most dramatic of which have since been named by the National Park Service....

A telegraphic synopsis for the second volume of Frémont's *Memoirs* notes the desperate sufferings of his party which were to begin here:

> Starvation again. Plenty of snow and no game.—THE OBELISKS.—*Cache* our baggage. Men and animals weak. I give out on mountainside—first time in all my journeying. Weakness temporary.

The spectacular steel engraving *Natural Obelisks* is the only vertical landscape composition among those remaining from the fifth expedition. It is the final surviving image, although a few more daguerreotypes were made before Carvalho's equipment was abandoned. The snow in which the expedition later foundered is apparent in the engraving, and its shadows indicate that the daguerreotype was made in the morning. From the site of [*Carvalho's image titled*] *Colorado Valley* (Wild Horse Butte), which is shown without snow, to the "Natural Obelisks" is about forty miles, which probably took Frémont's weakening men two days. The date of Carvalho's daguerreotype of the obelisks would then be January 23 or 24, 1854. About January 24 the expedition entered the Wasatch Mountains [*Thousand Lake Mountain*] by climbing out of Cathedral Valley at its westernmost extreme.

Kent Jackson [*a Wayne County man who managed the Capitol Reef orchards for many years*], who from his familiarity with this area immediately noticed that the engraving *Natural Obelisks* was laterally reversed, also demonstrated to me how Frémont must have skirted the Moroni Slopes and gone up Salt Wash into Cathedral Valley, passed by the "Cathedrals" and the obelisks at the north end of what is now Capitol Reef National Park, and then climbed out between Geyser and Hens Hole peaks to the west. Thanks to Jackson, Capitol Reef National Park now has a secure claim to a segment of the route of Frémont's fifth expedition.

When I was in the region, it looked as if a modern daguerreotype of the obelisks with the snow as is shown in the engraving was an impossibility, since the mild climate of the Great Colorado Valley has not changed since Frémont consulted on it with the Utah Indians in December 1853. However, the night before I went into the Middle Desert to daguerreotype the obelisks [*locally called "Mom, Pop, and Henry," names coined by Kent Jackson's father, Worthen Jackson*] the region was blanketed with snow, although by the time I reached the area and began work most of it on the slopes near the formation had disappeared. I was very fortunate, since the Indians were correct—snow here is rare and does not remain long....

The entire Cathedral Valley is a wonderland of endlessly varied sandstone formations. The "Monoliths" are just a few miles west of the "Natural Obelisks" and are the last of the great ones which Frémont and Carvalho would have passed before entering the Wasatch Mountains. There Frémont gave out for the first time in his life.... Later, he dealt with the incident more completely:

> We...were soon again involved among the snowfields of the mountains. There remained now only the bed of the Wahsatch range to cross. Here, for the first and only time in much travelling through inhospitable lands, I fairly gave out. Going up a long mountain slope, I was breaking my way through the snow a little way ahead of the party, when suddenly my strength gave out. All power of motion left me; I could not move a foot. The mountain slope was naked, but it just happened that near by was a good thick grove of aspens, and across a neighboring ravine the yellow grass showed above the snow on a south hillside. Saying to Godey [*a misremembered detail by Fremont; according to Robert Shlaer, this was almost certainly the Delaware Indian scout, Solomon Everett*], as he came up, that I would camp there, I sat down in the snow and waited. After a few moments, strength enough came back, and no one noticed what had happened. The next day we came upon a good camping ground, when I made a halt and disencumbered the party of everything not absolutely necessary.

...The abandonment of the heavy baggage spells the end of Carvalho's daguerreian activities with the expedition, but the daguerreotypes Carvalho produced were not included in the cache. Frémont retained these invaluable records and ultimately returned to New York City with them.

The cache itself would, if found, provide a wealth of information about the expedition and especially about Carvalho's methods, since it contained

(*above*) *Natural Obelisks*, steel engraving from a daguerreotype by Solomon Carvalho, the first photograph taken in Capitol Reef; courtesy the Huntington Library, San Marino, California. (*opposite*) Robert Shlaer's 1996 daguerreotype *Natural Obelisks*, re-creating Carvalho's original image from Cathedral Valley. Used with permission.

his equipment. Unfortunately, all evidence indicates the hoard was recovered in the spring of 1854 by a former member of the expedition who had been left by Frémont at Parowan. According to contemporary documents, this man was subsequently killed by Indians, who appropriated and disbursed the material. My attempts to locate fragments of Carvalho's apparatus in the museums of Utah have been unsuccessful.

Thus ended Carvalho's grand panorama of the West, which he made in the winter of 1853–54 for Col. John C. Frémont.

# 10

## PHOTOGRAPHED ALL THE BEST SCENERY (1872)

JACK HILLERS (1843–1925)

*In 1869, John Wesley Powell led the first scientific exploration of the Colorado River through the Utah canyon country and the Grand Canyon. When Powell needed another boatman for his second expedition in the spring of 1872, he hired Jack Hillers, a German immigrant working as a teamster in Salt Lake City when Powell encountered him. Hillers assisted Powell's photographers until his skill, stamina, and good humor (the boys called him "Jolly Jack") earned him the position of permanent field photographer for both the US Geological Survey and the Bureau of American Ethnology for the remainder of the nineteenth century.*

*This section of Hillers's journal documents the pioneering 1872 trek led by Powell's chief topographer, Almon Thompson, as he and his men, including Hillers, cross Boulder Mountain and the Waterpocket Fold, bound for the Henry Mountains and the Dirty Devil River. One of those newly explored peaks in the Henrys would later bear the name Mount Hillers.*

*For these Anglo newcomers, this really was exploration. Thompson and his field crew were working out the country. When they reached the Escalante River basin, it took them a while to realize that this wasn't the Dirty Devil River, that they had just discovered a major new stream. They relied on Utes for advice and local knowledge—but wrote about Indian people with the casual racism of men of their time.*

### June 11

In the afternoon we left the southern slope of the Lake Mountain Range [*Aquarius Plateau*] and crossed to the east side. The country below us is all cut up with gulches and cañons for miles—nothing but sand rock is visible.

The Dirty Devil Range [Henry Mountains] from this point looks like a dry country and almost impossible to get to them. The first peak is about 30 miles from this point eastward.

### June 12

Prof. [*Thompson*] and Capt. [*Captain Pardon Dodds, who had served as the Uintah Indian agent before working with Powell's men*] went in a northeasterly direction to find a trail or hunt up the best place for to get to the mountains near which our river flows. Fred [*Dellenbaugh, the young assistant topographer who had been through the Grand Canyon with Powell on his river trip earlier in 1872*] and Johnson [*William Derby Johnson, a Mormon settler working for Powell in 1872*] went southeast on a similar errand. Fen. [*Fennemore, the photographer*] and myself went down a cañon due east for pictures. Secured three—a storm coming up we returned to camp. Shot a dusky grouse; on our return found Fred and Johnson home. They reported having found a heavy beaten but not very old trail leading in the direction of the mts. About 5 p.m. Prof. and Capt. returned. Reported having seen fresh Indian sign. Rained very hard.

### June 13

Followed the trail F. and J. had found which led us through small valleys and cañons and to wind up, lead us down from the plateau on smooth sand rock over a thousand feet high—horses often slipped for a number of feet. Sometimes found ourselves on projected shelves, not over two and a half feet in width, but got down all OK. The last turn brought us in sight of a beautiful valley running northeast [Pleasant Creek] while from the west came a rushing stream which flowed in a small cañon of sand. All along the edges the cottonwood trees flourished. Camped in the cañon. Plenty feed for our animals—wild oats in abundance. Followed fresh Indian tracks all day.

### June 14

Up and off early. While making a descent from a Bench we were attracted by the bark of a dog. Looking in the direction from hence we heard it we saw two squaws flighing through the grease wood, yelling as though they had seen so many devils. We saw their camp on a small hill for which we

*Old Time rocks. Tantalus Creek, Aquarius Plateau, Wayne County, Utah.* Photograph by Jack Hillers (National Archives).

started. On nearing an old man about seventy met us at the foot of the hill, trembling like an aspen leaf. On reaching this camp found it deserted. Guns, bows and arrows had been hastily left in their wickiups, but we soon dismounted and seated ourselves around his camp fire, where we allayed his fears by telling him to smoke, at the same time handing him tobacco. At this token of friendship he steadied his nerves and began to talk. Found him to be one of the Red Lake Utes down here to gather seeds. A quarter of a hour after in came the two frightened squaws, who began to chatter like monkees. [A] little while after we saw two young men across a gulch on a small hill. The Squaws began to call to them not to be afraid for we were Tuitchea Ticabu (very friendly). They soon came slowly toward us looking like shamed men—no doubt they felt so. After smoking and talking, They begged us to go into camp and trade with them. Prof. being anxious to know about the country at large, and all about the trail, he consented to camp—traded several buck skins. I obtained a splendid skin for a small paper of paint.

## June 15

Struck off on a trail up the valley—followed it into a cañon. About a mile from its mouth we lost sight of it with the exception of a track here and there—probably cattle feeding. This is the great hiding [place] of the Indians and many heads have found their way in here, all stolen from the Mormons, who never suspected for a moment that their friends the Utes would do the like, but thinking the Navajoes the guilty Party. The Utes always watched the opportunity, when a band of the former were in the settlements. The Mormons could track their stock for quite a ways, but as soon as they got into a sandstone country they gave up the chase, being impossible to track them over the bare sandstone, and no one thought it for a moment possible to get down to the valley below. They never dreamed that here the Indian feasted on broiled steak. Wild oats grow here the same as cultivated does anywhere else but not so heavy. At present is the time when the Indian gathers his yearly supply of seeds and nuts. Those which we left had quite a crop gathered. I have no doubt that our party is the first white party here. The Indians felt surprised how we got in—asked numerous questions—how we found our way in. Wandered about all day trying to find a trail which would lead us out of the cañon besides the one we came in by without success. The cañon walls are perpendicular from 700 to 1000 ft. in height.

## June 16

On the search again. At noon Prof. concluded that four of us should climb out and head all the gulches. Prof. and myself went one way while Capt. and Fred went the other. Found a nice level Platteaux on top, but no sign of a trail, but found some very old horse dung, probably three or four years old. After heading all the side gulches we walked towards the mountains. Here we found ourselves on the divide. We could see the break of the Colorado plainly and only about twenty miles distance. Half the south side of the mountain is drained by a large cañon flowing east to c. [Colorado] while the other half flows west and then north into the Dirty Devil River which flows on the north side east into the Colorado. Having studied out our course we were bound to get out of the cañon. On our return we climbed down one of the side cañons and here we thought we could get up with a little work, but rather hard for the horses, but we were bound to get up. On our return about 4 PM found Capt. and Fred in camp, who reported having seen no sign of a trail, but had found about three miles from where we climbed out, two water pockets, and as our camp was about that distance from the water in the cañon, we concluded to try our cañon wall. After two hours' work we reached the top. One horse fell backward while going up a steep ledge and fell about ten feet, but picked himself up, shaked himself, and tryed it again—that time succeeded in reaching the top. Camped near the pockets.

11

## GILBERT'S NOTEBOOKS (1988)

### Charles B. Hunt (1906–1997)

*Charley Hunt, as his fellow scientists knew him, was a legendary field geologist known, especially, for his work on the Henry Mountains and the geologic history of the Colorado River. He cofounded the Utah Geological Society and taught at Johns Hopkins, the University of Utah, and New Mexico State University. He also served as director of the Geology Unit for the US Geological Survey. Hunt began his career in 1929, just eleven years after G. K. Gilbert's death, and he loved retracing Gilbert's footsteps.*

Gilbert's field notes about Utah's Henry Mountains are far more than a day-by-day account of his activities and observation. Reading them in the context of the status of knowledge about geological science at that time, 1875–1876, one quickly finds that the notes are an exciting record of what were totally new scientific discoveries representing milestones—breakthroughs—in the history of geological knowledge. One experiences the thrill of reading the firsthand account of the discovery of several concepts that today are accepted routinely as fundamental in the science.

Grove Karl Gilbert, 1844–1918, was one of the greatest geologists America has produced. Many share my opinion that he was the tallest of the several giants who contributed so much to the healthy development of the young science. His contemporaries evidently thought so also, because twice they elected him President of the Geological Society of America, the only person to be so honored. His contributions, of course, are recorded in his published

Henry Mountains from above Fivemile Wash, Capitol Reef, 2012.

monographs, but his field notes provide an inside view of how his major discoveries developed.

## Why Gilbert studied the Henry Mountains

John Wesley Powell, in his trips down the Colorado River (1869–1871), had a good view of the two southern Henry Mountains and saw that they were structural domes associated with "lavas." Geologists of that time still were debating whether volcanoes were "craters of elevation" [*formed by the cataclysmic upheaval of overlying strata*] or merely piles of accumulated lava and other ejecta around the craters. Powell could see that the two mountains were domed, for the massive Triassic-Jurassic sandstones conspicuously rise on their flanks. The "lavas" could be seen as dark masses at the centers of the domes, and specimens could be examined in the drainage courses off the mountains. Surely these were craters of elevation.

And so Powell arranged for Gilbert to visit the Henry Mountains and determine the facts. To reach the Henry Mountains, Gilbert had to cross the lavas of the High Plateaus, from Salina Canyon south to Fishlake and around Rabbit Valley east of Fishlake. From a distance, and for quite a while after being in the Henry Mountains, Gilbert continued to accept the idea that likewise the igneous rocks in the Henry Mountains are lavas and referred to them as such in his first notebooks.

## Getting there

Most of southeastern Utah and all of the Henry Mountains area was a vast, unmapped, vacant wilderness during the 1870s when Gilbert made his trip there. Of the Spaniards who explored the southern part of the Colorado Plateau in the 16th century, only Cardenas saw the Colorado River—at the Grand Canyon in 1540. Only Escalante, in 1776, crossed the river. Ives (1861) was the first of the U.S. exploratory surveys to describe the region. He wrote discouragingly, "It seems destined by nature that the Colorado River, along the greater part of its lonely and majestic way, shall be forever unvisited and undisturbed." This still seems to be the view of Utahns, despite the fact that southeastern Utah's unparalleled scenery is potentially the state's greatest resource.

Other exploratory surveys east of the Colorado River while Gilbert was visiting the Henry Mountains could have seen the Henry Mountains only as unidentified peaks. Holmes' [*W. H. Holmes, another of the Powell Survey's scientists*] interpretation of the intrusive structure at Ute Mountain (Sierra El Late) anticipated Gilbert's interpretation of the Henry Mountains. In fact, Holmes even recognized that the bulging intrusions there are linear and radiate from a center. And Gilbert surely was mindful of the fact that Peale's [*Albert Peale, yet another nineteenth-century USGS geologist*] attempt to study the La Sal Mountains had to be aborted because of hostility of the Indians.

The Henry Mountains did not appear on any map until Powell discovered them in 1869. Powell named them for Joseph Henry, physicist at the Smithsonian Institution who assisted Powell in obtaining financial support for his exploration of the Colorado River. Even as recently as the 1930s, when we made our survey of the region, it still was the center of an area the size of New York State without a railroad (and it still is), and a third of that was without a road of any kind. The Henry Mountains and the country around them still was pack-train country, and to work there provided an opportunity to share the experiences of pioneers like Gilbert and his party who made the first maps. Ours was the last of the big pack-train surveying projects that had begun in the 1850s as surveys for a transcontinental railroad—the end of an era.

How did Gilbert find his way across thousands of square miles of vacant, unmapped, rugged wilderness to reach the Henry Mountains, and how was he able to always complete a day of travel and end at a spring? He did so by following the prehistoric Indian trails. So also did the frontiersmen who explored for grazing or farm lands and the three parties of prospectors who had preceded him to the Henry Mountains. To explorers like Gilbert, those Indian trails were as easy to follow as a modern freeway. They were the conspicuous routes through the otherwise trackless wilderness before there was grazing by livestock. Moreover, one can even find directional signs along the trails indicating whether one is approaching water or going away from it. The secondary trails of coyotes, foxes, and rabbits diverge from the main trail away from water; they converge toward water. These signs of the past still are preserved in Death Valley where there has been no grazing. That Gilbert knew these signs and could use them to his advantage is indicated by the ease with which he generally found satisfactory camp sites that provided the pasturage and water for his animals as well as his men. In two field seasons, only a couple of his camps proved unsatisfactory.

To considerable degree this still was true of the Henry Mountains area during the 1930s, and the geologists had to learn to distinguish between the "big" trails and the "dim" ones. The ancient Indian trails had by then been destroyed by livestock, but the cows and wild horses marked the routes to and from water.

Gilbert knew he was pioneering in exploring the Henry Mountains area, and his published report begins: "The Henry Mountains have been visited only by the explorer." He further wrote, "No one but a geologist will ever profitably seek out the Henry Mountains…," but he reveals his enthusiasm for the area by following this statement with three or four pages of discussion of how to get there. "At Salt Lake City he can procure pack-mules and pack saddles, or 'aparejos' and everything necessary for a mountain outfit."

# 12

## IN CATHEDRAL VALLEY (1915)

### George C. Fraser (1872–1935)

*Dave Rust (1874–1963) combined a childhood in southern Utah, an education at Brigham Young University and Stanford, and a boundless love for the natural history and geography of the Plateau. Working with this mix, he pioneered the role of today's backcountry guide. Educator, elected official, entrepreneur, poet, cowboy, devout Mormon, gardener, and guide, Rust hoped those who adventured with him would "love my country—Powell loved it; Dutton loved it; I love it, and so must you."*

*Each summer he led his well-heeled travelers from the East Coast on month-long pack trips and river trips through Capitol Reef, to the top of Boulder Mountain, and across the southern Utah canyons. Here's a description of one of those trips, from George Fraser's journals, edited by Fred Swanson—who is Dave Rust's biographer, as well.*

*Fraser made a half-dozen trips to the Southwest over twenty years, with his son or one of his daughters, in turn. Fraser was a Wall Street lawyer with a degree in geology from Princeton and an unending curiosity about both the land and people of the Southwest. Dave Rust was his favorite guide. Fraser was Rust's favorite customer.*

### Tuesday, July 6, 1915

The South and Middle Deserts are valleys with steep cliffs bounding a flat waterless sandy bottom. Our camp was at the summit of the cliff forming the terminus of the South Desert approximately 400 feet high.

[*Dave*] Rust had been familiar with this country when a boy and up to about 25 years ago, but since that time had not been here. Nevertheless, he remembered where water was to be had. While he and George [*the writer's son*] took the water keg and water bags about a quarter mile northerly from the camp, I sat on the brink of the cliff bounding the South Desert and studied the Henry Mountains and the water pocket fold, the latter forming the southerly or southwesterly boundary of the South Desert. After getting water at 6:30 p.m. we walked about three-quarters of a mile to the head of Middle Desert. The coloring and sculpture here is beyond anything we have seen. The cliffs rise precipitously from the flat, barren desert floor 250 feet, perhaps 350 feet. They are bedded horizontally. The top is of white, very soft cross-bedded sandstone, beneath which is a brilliant carmine sandstone or shale. Rising from the desert floor separate from bounding cliffs are numerous temples. These are wholly of carmine rock, the white bed of the summit having been eroded off. The desert is sandy, cream to yellow, with a few piñons and cedars showing patches of green. On the summit of the cliffs, accentuated and silhouetted as it were, basalt boulders lie on the pure white sandstone.

The sun set clear save for some fleecy clouds to the north. Sitting on the brink of the cliff at the head of the South Desert, the Henry Mountains rose against the sky, appearing in the light of the setting sun first blue, then grayish in the haze of the cloudless sky, and as the sun went below the horizon and still touched them, they appeared burnished with gold which gradually, as the light waned, turned to lilac until their illumination ceased.

The prevailing tone of the foreground was magenta with white and terracotta streaks due to the wash from above. In the desert were patches of piñons and cedar, the distant ones appearing black, those nearby vivid green. The silence was complete save for an occasional bird call and the buzzing of gnats and mosquitoes. These were very troublesome until sun-down, but did not bother us thereafter. The change in temperature this evening from last night was very marked, but the night was comfortably cool.

### Wednesday, July 7, 1915

Up 5 a.m. Clear, except for a few clouds to the north. A golden sunrise.

Left Camp 8 [*at*] 7:15 a.m. The horses had poor pickings all night and we had a long ride to go to Caineville over a dry and desolate route. Rust told me that when ten years of age he was once taken over this desert and some time afterward was told to take his horse and a pack animal and make alone the trip we were setting out on. He started early and long after dark found his way back to camp. The pack had turned on his horse. He did not know how to pack the animal himself, he did not know his way and all he could do was to get himself back to the point from which he had started. That is the school in which he was brought up and where he learned the secret of direction and how to fix land-marks.

About four miles down the valley we saw rising from the valley floor a curious three-fingered butte or temple, a remnant of a lava dike that crosses the valley. There are many of these dikes. Some of them can be traced for a mile or more; also some lava flows capping the surface of the plateau on the easterly side. A large butte rises in the center of the valley, and a mile beyond that appears a graceful Temple, the most picturesque feature of the landscape on this trip. This rises about 400' from the valley floor and is composed of magenta sandstone or shale, the overlying beds having been eroded off.

It was hot and the dust raised by the pack horses which we drove made riding trying. As there was no water between the spring and Caineville, we filled the keg and loaded it on the mule. About 11 o'clock the mule tried to roll, the water barrel having chafed its back, so we stopped for three-quarters

Lower South Desert, Capitol Reef, 1977.

of an hour, repacked the mule, drank all the water we could, emptied the barrel and ate a little. The traveling in a southwest wind was hot, but the sun's rays were tempered by a few clouds.

About two miles beyond where we repacked the mule, we turned southeast up a draw and left South Desert, descending into the two-mile wide Blue Flat. The bottom of Blue Flat is a dry, powdery grayish mud sun-dried and in low spots like a quick-sand, very treacherous, so that we were compelled to keep to high ground. It is entirely barren and very dusty.

Ascending the easterly up-dip slope of Blue Flat over bare rock we came to the summit of a precipitous wall bounding an anticlinal valley known as the Red Desert. This is an elliptical area bounded by a wall of reddish shale apparently only broken down at one point. The bottom of this valley is uneven and hummocky. The red beds are interspersed with whitish and bluish gray thin layers of shale entirely bare of vegetation and bedded nearly horizontally. On account of this coloring and the uneven weathering of the surface, the floor of the desert is striped and blotched and patched like a cross between a leopard and a zebra.

We climbed the wall to the south, first over the red rocks and then over a yellowish sandstone and conglomerate, followed a deep wash through the yellow sandstone southwesterly into the valley of the Dirty Devil [*today's mapmakers call this river the Fremont until it joins the Muddy River at Hanksville to form the Dirty Devil*], striking the road running parallel with the river about one mile above Caineville. Arrived 3:15 p.m. Very hot. We camped by the road at Curtis's Ranch, the first one we struck on reaching town, on the north bank of the river. We were very tired, hot and dirty, so at once made for the river to get a bath.

The Dirty Devil is full of silt, very shallow and very swift. George and I undressed on a mud-bank in full sight of the whole town, but there seemed to be no one there that we could disturb. We lay in the dirty water and rolled over on the pebbles, with each roll getting a little more of ourselves wet, and all the while fought mosquitoes that covered us like a blanket. If not cleansing, the bath was at least cooling, so that we were able to partially enjoy a belated lunch around 5:30.

After lunch or supper we walked "down town." The town consists of a few, perhaps 15, tumble-down shacks strung along the north bank of the

Dirty Devil over a distance of a mile and a half. We asked a boy where the store was. He said Osterberg (a Swede or Dane) kept a little candy but there was no store. We found Osterberg's and were told that everything he had, which consisted solely of candy and tobacco, had been sold out on the 4th of July, except one box of chocolates which sold for $1.00 a pound and that apparently had been there for years. We induced Mrs. Osterberg, however, to let us have five pounds of sugar from her private stock.

# 13

## ONE OF THE GREAT TRIPS OF THE WORLD (1922)

### John A. Widtsoe (1872–1952)

*In the fall of 1922, John Widtsoe, LDS apostle, Harvard-educated chemist, agricultural reclamation scientist, and former University of Utah president, floated through Glen Canyon with a group of engineers and high-level bureaucrats to survey dam sites. They came here to prepare for the upcoming November meeting in Santa Fe, New Mexico, to portion out the Colorado River's water in the Colorado River Compact. To reach Glen Canyon, the party passed through what would become Capitol Reef National Park, via the "road" through Capitol Gorge, then south all the way down Halls Creek to the Colorado River.*

*Widtsoe's journal wasn't published until after his death, but he turns out to have a dry wit and sharp eye. His journey resonates with the integrity of the Waterpocket Fold before Glen Canyon Dam, when Halls Creek flowed unfettered to the Colorado River—a connection lost beneath the reservoir planned so enthusiastically by Widtsoe and his companions in their certainty that the Colorado River needed "controlling" and water needed to be "stored."*

### Sept. 4th, Monday

Left Richfield about 8:30 a.m. Delay due to one machine breaking. One big truck for equipment and three of our party. Two passenger autos for the remainder. Drove by way of Glenwood over mountains. Clouds in great masses under a blue sky. Stetson wonders if we shall see anything finer. Caldwell doesn't know. Everybody else has opinions. Drove to Grass Valley over mountain—9,200 feet elevation, down to Loa. The day again perfect, rather cold....

Left Loa at 1:30 p.m.... In short time the high plateau country opened with its variety of colors, fantastic shapes; aeolian cuttings, buttes, towers, etc. Passed through many little villages—Lyman, Bicknell, near Teasdale, Fruita, where we had fruit. In Fruita met party of young ladies well dressed which caused much favorable comment from the party.

Heavy rains yesterday had cut into Capitol Wash roads. About four miles down road had to be repaired 6 or 7 times. Whole company out making roads—cars swinging at all angles over broken roads. We got through without mishap. A glorious canyon—approaching at the upper end, Zion Canyon. The day perfect with a few threatening clouds. Caldwell, Stetson and Dennis, on freight wagon, up in the air part of the time. Looked like professional acrobats. Stetson takes photos from top of truck—probably moving pictures. Everybody happy.

Out of Wash a mile or two darkness came. Camped near first ranch on Pleasant Creek [*Notom*]. Caldwell, Stetson and I made beds under a branching cottonwood, not far from Pleasant Creek. It rained, or rather sprinkled. Too dark to investigate larder. So bacon and eggs. Butter lost. Full moon—fine evening—Stetson sore and tired. Talks of sleeping. Elev. 5800 feet. LaRue calm, but terribly worried about the butter. A.P. Davis during day told of organization of Geological Survey. We decided (?) to rename Capitol Wash—*Capitol Gorge.*

### Sept. 5, Tuesday

Broke camp at sunrise and left Pleasant Creek. Still in autos. 8:30 a.m. reached Bowns Ranch on Sandy Creek [*today's Sandy Ranch*]. A shack or two—headquarters for the ranging crew. Tough automobile roads. Saw two wild animals and squirrels on the trip! The scenery beggars description....

Changed baggage here from auto to wagons. Two four horse teams. Baggage and bedding and provisions—in wagon; passengers on top. A hot day, with a gentle wind moderating the heat. No shade. Drove at about 3 miles per hour.... Everybody in good temper. The scenery continues wonderful. We travel in gulches and draws. Vegetation very scarce, though at Bowns an abundance of Russian Thistle.... Unusual range of layered rocks, towering, 1000 feet high on one side and equally high red sandstone cliffs on other. Towers, castles and buttes etc., break the level strata.... Towards evening left red sandstones, into a narrow valley of white sandstone and limestone. No water; no animal life; one mourning dove all day; scanty vegetation; true desert. Here and there, where a flood has washed, there are deep chasms showing soils 20–30 feet deep. Traveled from Bowns about 18 miles.

Camped in a dry spot by side of road, near Bitter Creek. Made beds; then supper. Party tired. Our camp is in a valley about 1½ miles long; ¾ miles wide. The usual horizontal white and red rocks seem tilted on end. On the side of the valley opposite to the camp east runs *Bitter Creek*—which is not much more than a seep. Over the low ridge west of the camp is a small romantic valley; which, as the full moon filled it with mystery and beauty held me for a long time. On the east ridge is the clear outline of a camel resting. The head is distinct; the neck somewhat depressed, the back with two humps very marked. This is a good land mark. Camp and place called *The Camel's Rest.*

### Wednesday, Sept. 6, 1922

Full moon throughout the night. Warm and balmy until early morning when a wind arose followed by a drop in temperature. All up at 5:30. Breakfast. No water for washing unless we go nearly a mile to Bitter Creek. An unwashed crowd. Everybody happy. Stetson did not sleep well. Mr. Davis loaned me needle and thread and I sewed on button.

The rising sun lighted up the yellow west wall with a golden splendor, glorious to the soul. A flock of geese cackled through the gray of the dawn. Some mourning doves were seen at sun rise. Crickets here and there faintly throughout the night. We were off for Muley Twist. No one knows distances from Bown's ranch. Probably about 60 miles to River. The draw narrows and the scenery becomes rougher and more picturesque. About six miles down we pass the *Crotch* where a branch of the road takes off into Hanksville country. Over hill and dale, skidding on rock and soft earth and making sharp curves we finally come to Muley Twist Canyon (a muley horse died there.) The road through Muley Twist Canyon takes off from our road and goes into the Escalante, Hole in the Rock, and Tropic (Bryce Canyon) country. Muley Water Holes are just south of the Canyon. The whole side of the gently sloping mountain is white sandstones with pockets here and there. Water collects in these holes. The lower ones accessible to stock, the higher ones are not. On the east side of Muley are the red sandstones. One huge dominating red butte, the largest so far, forms a decided landmark. We are traveling in Grand Gulch. Numerous places here could well be named. Three miles above Muley on west of road is a great red sandstone bluff, which from the north looks like a *sphinx*. We named it Sphinx Bluff. A little lower down on the very edge of the western cliff is the outline of a locomotive, furnace and all. We named it *Locomotive Valley.* We stopped about 11 a.m. at Muley Tanks for lunch. We had lemonade! Also good water!!

...Some cattle seen along road this a.m. Nearly all white faced; sleek and fat. LaRue and Birdseye hunted cotton tail rabbits. Shot four. There is so much magnificent scenery that in spite of our best desires, we are not appreciating it as we should. Marvelous how man becomes callous! After lunch I picked flints and agates from creek bed and chips from Indian work. Also one spearhead and a beautiful dark red flint knife (skinning knife.)

We traveled down Grand Gulch [*Halls Creek*] on the rocky creek bed. The Gulch has narrowed, the scenery has increased in grandeur. About four p.m. we stopped for fifteen minutes at Fountain Tanks hollowed out by wind

Lower Muley Twist Canyon at sunset, Capitol Reef, 1975.

in spots of soft material. Then on to head of Hall Creek where we camped for the night. At this point on west side of canyon, great sloping mountain sides of gray sandstone or limestone. Pots hollowed out everywhere. Pools of water in river or creek bed, though evidences everywhere that at flood time the water runs 14 to 20 feet or more higher. In one case a cottonwood stands exposed in six feet of earth, the bark fully down to bottom, showing how earth had been gathered around tree, then washed away again. A fluctuating country wherever the water can reach it.

About three miles up from camp passed some high cliffs composed of earth and pebble conglomerate. Weathering wherever a flat rock had fallen had produced high pillars, with the rock on top. *Named it Toadstool Curve.* Fine moon again at night though late because of 1000 feet of vertical red cliff just beyond our beds, on east side. Fried rabbit for supper. In bed about 9:30 p.m. Camp elev. 3878.

### Thursday, Sept. 7, 1922

Up at 5 a.m. Started at 6:30 a.m. Crossed Hall's Divide. Bad road. Then down creek bed nearly all the time until we reached Hall's Ranch at 10:15 a.m. Half of party went through canyon of Hall Creek [*Halls Creek Narrows*]. Wonderful they declare. Called it Canopy Gorge from overhanging rock. Vast amphitheatre there with perfect acoustics. About six miles of rounding narrow gorge—in places only 30 feet across. Rested horses and had lunch at Hall's Ranch. Old ranch house now moved away. Country continues the same. High horizontal red cliffs to east; high nearly vertical gray cliffs to west. Probably traveling along fault line, now Grand Gulch. Party that went through Canopy Gorge doctoring blistered feet.

This whole trip from Loa down has been wonderfully picturesque—more than that: full of magnificent scenery, and a remarkable variety of forms. A good trip for a geological party or for any lovers of nature. The country is of course similar to that on the road down to Cainsville and Hanksville, but very different in detail. To go down this way, then up the river and back by way of Hanksville would be one of the great trips of the world. Publicity should be given to this region. Hall's Crossing named from Hall's creek which enters the Colorado River at that point. Hall's Creek named for Hall's Ranch, at which I am now writing. We have traveled along one of the emigrant trails leading to San Juan County. Mr. Pace, one of the original party to cross the Colorado in 1879 still lives in Loa, where I had a conversation with him as we came down on the trip. All along the road in Grand Gulch, upper portion especially, are small areas that could be cultivated if water were available. Apparently from the flood marks, there is ample water leaving the gulch to care for all the lands and surplus, a big one, if stored. There are numerous places where storage could be effected; the chief danger and obstacle being the shifting conditions at flood time, and the fierceness of the floods when they come. The reclamation of these small areas is not, probably, a problem of this generation....

Left Hall's Ranch for Baker's Ranch at 12 p.m. Reached Baker's Ranch at 1 p.m. Small three roomed shanty, cistern and corral at Bakers. Now run by two Baker brothers, sons of Brother Baker of Richfield who bought the ranch. Left Baker's at 1:20 p.m. Canyon widens. Can see across Colorado into San Juan Country....

The cliffs or mountains that have followed us on the east all the way down Grand Gulch—great thicknesses of chocolate colored rock, in horizontal strata, underlaid by red rock—terminates a mile or two below Bakers. A bold magnificent cliff projects at the end looking towards the Colorado River, and forms a great landmark. Lower down towards the Colorado River we can look back and see the other side of this chocolate capped red rock beneath the mountain. It is as sheer on that side as on the side we have seen from Grand Gulch. The mountain on the west side of Grand Gulch, gray and vertical, also terminates long before the River is reached. As we look back on them they are covered forest like with knobs or nipples. The party suggested we call it Nipple Mountain. On the way we get a magnificent view to the east of the Henry Mountains rising out of the broken country. The one farthest north seems the highest—probably Mt. Ellen. Look up.

...Hall's creek has cut deep into the rock, forming a very high and narrow canyon. As we approach the River, two miles away can be seen the sheer, red wall against which the River runs on the east side. Then, great, barren, smooth rocks like those on the beaten shoreline, except a hundred times larger, along which we slide. Suddenly, in a cut in the rocks, we get a view of the River. A long and wide beach, with several fine clumps of trees. Just beyond the beach down stream the River turns sharply to the left, just below the junction is Hall's Creek, between sheer cliffs, 600 feet high, and only 175 feet apart. All along the River, the cliffs to the last sheer and high red sandstone.

On the top of the first hill we meet the boatmen.... Over the cliffs we reach the camp in a beautiful grove of cottonwoods. I go down to look at the River and the boats. Evidences here and there of dredging operation and oil service. The Colorado—yellowish brown, but stately and certain of its own value. Relatively few men have been here. Many emotions arise. The wagons arrive. All well. The first stage of the expedition is over. Just five days from home.... Early supper, cooked by the boatmen. Talk and plans. In bed at 9:30. The full moon filled the valley with a new beauty. The high sheer wall to the east looked dark and forbidding; the lower range on the west glowed with a living glory. The small valley had its sounds—the sounds of the desert—intense in their stillness. God be thanked for the earth and its beauty; for life; for hope—and now at this moment for our safe journey so far.

The Henry Mountains and Tarantula Mesa, from the Waterpocket Fold, 2012.

# PART IV

## CANYON COUNTRY

### Earth Unmasked

The Capitol Reef landscape challenges and confounds. All that rock, all those singular plants growing where no plant should find life easy, all those animals mostly active at night whose lives we rarely encounter.

And so we need the insights and voices of scientists and naturalist writers to help us understand. Ward Roylance spent his life trying to capture the Colorado Plateau in words. I include his introduction to his Capitol Reef guidebook here. Then there is Abbey.

Edward Abbey remains the voice of the canyon country thirty years after his death in 1989 and a half century after the publication of *Desert Solitaire*, his extended essay and elegy for Utah's redrock wilderness. The short piece here, written as the foreword to a 1986 book of photographs that I edited, *Blessed by Light: Visions of the Colorado Plateau*, is filled with Abbey's characteristic lyricism, anger, natural history, philosophy, call to action, and self-deprecatory humor. He was one of a kind.

Starting with the foundational nineteenth-century surveys of Lieutenant George Wheeler and Major John Wesley Powell, geologists have made the Colorado Plateau a place where fieldwork predictably yields discovery. The nature of their work, immersed in deep time, tends to make them comfortable with pondering the extraordinary.

Writer Charles Bowden revisits Wheeler's work and does his best to wrap his imagination around the immensity of geologic time. Writer and photographer Michael Collier brings us the modern geologist's voice in an excerpt from his book about Capitol Reef geology, written for the park visitor.

Natural history writers make crucial companions for us all when we enter new country. These writers have read the science. As relentlessly observant wanderers, they have soaked up the spirit of the land and its plants and creatures—and then turned their experiences into rich and surprising language. Here, two of the best naturalist writers, Ellen Meloy and Ann Zwinger—the latter called herself with wry self-deprecation "the near-sighted naturalist"—give us their take on Capitol Reef wildflowers, aspen, and toads. Last, Ron Lanner, forest biologist and science writer, shares his enthusiasm for the remarkable coevolved partnership of piñon pines and piñon jays.

Spend time with the words of these keen observers, and you'll never again see the canyons of Capitol Reef as monotonous or barren or "timeless." This redrock canyon country is dynamic and alive.

# 14

## AN INTRODUCTION TO THE PARK (1979)

WARD J. ROYLANCE (1920–1993)

*Ward Roylance began exploring the canyon country from Salt Lake City in the 1940s, and after decades as a travel writer he and his wife, Gloria, moved to Torrey in 1976. In the years before his death in 1993, Roylance wrote from his home, shifting from post–World War II boosterism to concern over the Big Build-Up of development in the Four Corners. He published a guide to Capitol Reef (excerpted here) and* The Enchanted Wilderness, *a love letter to the Colorado Plateau. His Torrey home is now a nonprofit arts and education center, the Entrada Institute.*

As the essence of a mystic experience—even a taste or smell—cannot be conveyed in words, so it is with Capitol Reef National Park. Its uniqueness defies language.

Geologically the Waterpocket Fold, foundation of the park, is not too unusual in such basic attributes as its rock layers (which are common throughout much of the Colorado Plateau), or its numberless erosional forms (which have general counterparts elsewhere), or its dramatic monoclinal structure (other picturesque monoclines are in the vicinity). Even the varied colors of its rocks, while superlative, can be found in other parts of the Four Corners region.

No, the uniqueness of Capitol Reef National Park lies not merely in its basic geological constituents, however remarkable they might be. The secret of its sublimity is in the blending, the polishing, the finishing.

The Castle and the Henry Mountains, from Panorama Point, 2015.

Any work of art requires raw materials. What transforms materials into art is some creative force, be it human genius or natural impulse, as well as a fortuitous combination of those materials.

In the case of Capitol Reef the creative impulse (erosion) has combined with ideal materials to create a masterpiece of *natural esthetics*.

Esthetics: *pertaining to beauty or the beautiful.* Perhaps esthetics is the key concept in evaluating the park. Its beauty is different from that of any other place. The park stands in an esthetic class by itself, setting its own distinctive standard of beauty. There is nothing else like it.

At Capitol Reef nature has taken ordinary raw materials (rocks), painted in a spectrum of rainbow hues, and shaped them into a work of amazing erosional intricacy. Maze, tangle and labyrinth are descriptive terms that are eminently suitable here. So are adjectives such as strange, marvelous, exquisite, harmonious, majestic—the list could go on without exaggeration.

And what to call the park's natural forms? The English language is hardly adequate to describe some of them.

**Canyons?** Yes, but not merely "canyons." In Capitol Reef they are chasms... gorges... gulches... slots... natural phenomena that Webster never dreamed of.

**Buttes?** Yes, but in Capitol Reef a "butte" is likely to be a rounded, swelling, doming, swirling object of esthetic perfection. The angular butte named by Spanish explorers bears little resemblance to the splendid butte-forms of Capitol Reef.

**Cliffs?** Yes, hundreds of miles of them. But in Capitol Reef they are not merely "cliffs." Here they have a delightful personality all their own, the result of rainbow coloring and wonderful erosional designs.

All of this color, erosional intricacy and uniqueness of personality make Capitol Reef a park for connoisseurs. There is no end to esthetic discovery. Every change in perspective reveals new forms and relationships. Clouds... sunlight... moonlight... time of day... rain... snow: every nuance of lighting and contrast creates new colors, shapes, spatial dimensions. The possibilities for visual experience seem boundless here.

# 15

## COME ON IN (1986)

### EDWARD ABBEY (1927–1989)

*With* Desert Solitaire *(1968) and* The Monkey Wrench Gang *(1975), Edward Abbey became the literary voice of the canyon country. His ornery prose demanded that the reader react, and not everyone loved him. But his followers (and his fans truly were more followers than mere readers) made him the muse of direct-action environmentalism, an Old Testament prophet thundering in defense of wilderness. He called his essays "antidotes to despair."*

*Nearly everything Abbey wrote grew from the redrock wilderness. This piece, Abbey's introduction to* Blessed by Light: Visions of the Colorado Plateau, *distills his voice.*

It is difficult to write about the landscape of the American West without lurching into rhapsody, as the testimonials in this book naively demonstrate. And probably no part of that landscape has provoked more rhapsodical prose, purple as the bloom of the sage, than what geographers call the Colorado Plateau province—i.e., the region drained by the Colorado River. For example, any author capable of typing such a phrase as "... the light that never was is here, now, in the storm-sculptured gorge of the Escalante," [*Abbey's own words*] should probably have his typewriter confiscated, hammered flat, and sunk in the silty depths of Lake Powell National Sewage Lagoon. With the author's neck attached by a short length of anchor chain.

This straining after mystical, visionary, transcendental typing has long been a source of amusement to critics and book reviewers among our East Coast literati. (That little clique and claque of prep-school playmates and

Ivy League colleagues.) In the minds of those stern-disciplined fellows, with their faces still set toward New England, Old England, central France and the mud of Mississippi, the extravagant language employed by Western writers when writing about the American West reveals a softening of intellect, brain cells addled by too much sun, the lingering afterglow from the sunset of nineteenth century romanticism.

Could be. Maybe they know what they're talking about. But they don't know what we are talking about. We, that is, who live in the West, love the West, hate the West, and persist in trying to paint it, photograph it, and describe its disturbing shapes, colors, and nature through the inadequate vocabulary of a language that was formed among the vapid bogs and insipid fuzzy hills of little England.

It cannot be done. The world is bigger than we are, fundamentally mysterious—why should *anything* exist?—and beyond the scope of human expression. No system of symbols can be expected to comprehend and apprehend that which is the source of all systems. Nevertheless we keep trying, groping, searching, feeling our way toward some sort of vision, some kind of understanding. In the process we are forced to expand the powers of imagination and stretch to breaking point the modes of communication. We always fail, we inevitably look ridiculous. The most heroic efforts lead only to another painting, another photograph, one more book. The best of these refer, finally, only to themselves.

Art cannot replicate the natural world—neither the starry universe nor the voodoo gulches and hoodoo rocks of the canyon country, the high mesas, the sunken rivers, the corroded cliffs of the Colorado Plateau. But these failures do not matter. Not if they result in books worth reading, paintings worth hanging on a wall, or photographs that command more than a caption, more than one passive look. The purpose of art is not replication but creation—the making of integrated little worlds within the greater world that encloses us.

Given that understanding, we can proceed to my second argument. It is not enough to describe the world of nature; the point is to preserve it. It is not enough to paint, photograph, or even to understand the American West; the point is to save it. It is not enough to admire or love the Colorado Plateau region; the point is to defend it from its enemies. Landscape aesthetes

Halls Creek, the most remote country in the park, 2011.

are common as lichens on an academic wall. What we need are heroes and heroines—about a million of them—willing and able to fight for the health of the land and its native inhabitants.

The world of Glen Canyon, now buried under the sludge and stagnant waters of Lake Powell, will serve as one good example of what happens when we compromise away too much. The Mojave Desert of southern California—smogged over by urbanization, bombarded by the military, devastated by weekend hordes of mechanical recreationists—is another. I'm hardly thinking of "wilderness," which barely survives in the forty-eight states, but simply of back country, primitive country, of land not yet wholly surrendered to industrialism and industrial tourism. I mean country like the Colorado Plateau, with its many millions of acres of still roadless territory.

Difficulty of access is the most democratic of screening devices. Anyone with the price of an old pickup truck in his bank account is free to explore and enjoy the back country of Utah, Arizona, the Four Corners—few things could be simpler, easier, cheaper. Dirt roads "lock out" only those people who fear dust, who cannot cope with a flat tire, who lack the sense to stash a little food, water, and a bedroll in the cargo space of the machine they drive and are driven by—those in short who lack, not money, but *interest*.

The same principle applies to what we loosely call "wilderness," meaning an area without permanent vehicular roads. Any man or woman in normal health, with the price of a pair of walking shoes in his or her pocket, is free to enter, to wander through, to get lost in, and to escape from the easy-going wilderness of the canyonlands. This form of mild adventure has become so popular now that in many places (such as Yosemite, Grand Gulch, the Grand Canyon) a walk in the wilds is subject to permits, advance reservations—rationing.

Which means, to my mind, that we will soon need far more, not less, official wilderness in our United States. More wilderness or fewer people. Preferably both.

The human animal needs adventure. Fantasy is not enough. But more important than our need for open space, physical freedom, occasional solitude, is the need for the land to be let alone and left alone. For its own sake. *Let being be*, said the philosopher Martin Heidegger. I can imagine a civilization wise and generous enough to set aside vast tracts of land where, by mutual agreement, none of us humans enter at all. Ever. A place where our feathered, furred and scaly-skinned cousins—as well as the plants and the rocks—are free to work out their evolutionary destinies without any meddling or "management" whatsoever.

Another fantasy, no doubt. Meanwhile, we must begin where we are, with our backs to the wall. Somehow, now or never, we must draw a line before the advance of commercial greed, industrial expansion and population increase and announce, in plain language, "Enough is enough. Thus far and no farther. Think of your children. Of their children. Of the hawks, buzzards, lizards, bear. Save a little room and time for the free play of the human spirit and the wild play of the animal kingdom."

I can think of no better place to draw that line—in words of flame, in deeds of conviction—than around the red rock, the sunburnt canyons, the lonesome junipers and the solitary mountain lion of the Colorado Plateau.

Come on in and see for yourself.

# 16

## IMPRACTICABLE RIDGES (1996)

CHARLES BOWDEN (1945–2014)

*Chuck Bowden's writing ranged from street reporting in Chicago to a string of harrowing books about the drug trade along the Mexican border to fierce and quirky natural history writing about the Southwest. This excerpt comes from his collaboration with photographer Jack Dykinga,* Stone Canyons of the Colorado Plateau, *which grew from time in the Escalante and Paria canyons.*

*Bowden focuses on the Wheeler Survey here, which overlapped with John Wesley Powell's survey at Capitol Reef in the early 1870s. Some men worked for both, in sequence, including G. K. Gilbert and geologist Edwin Howell. In his journals, Gilbert even refers to Capitol Reef as "Howell's Fold."*

Lt. George Wheeler, West Point class of '66, is probing the edges of the canyon and plateau country in 1869. He is part of what will become an invasion of scientific Vandals and Visigoths—the expeditions of King, Powell, Hayden, and of course, Wheeler. Everything here is very old, and for the men of science all this antiquity of stone is very new. It will not be until the early 1880s that they will agree on a simple language for their discoveries, the terms we now use, such as Quaternary, Tertiary, Cretaceous, Jurassic, and so forth.

Wheeler is smitten, and in 1871 returns with a passel of geologists and topographers, troops, a photographer, and, of course, a reporter. He is to look for minerals and coal and anything that indicates money can be made. He paddles upstream in the Grand Canyon and travels but fifty-three miles in twenty-four days. His ego is good-sized. Two years after Major Powell and his men have raced down the entire canyon country of the Colorado, Wheeler announces of his midget foray, "The exploration of the Colorado River may now be considered complete."

He is part of a breed. While the Saints are putting their faith and lives on the line to settle their Zion, the scientists are racing each other to cover as much country as possible—hustling like lobbyists on Capitol Hill, pitching their wares in the newspapers, sniping at each other often through the various bureaucracies and agencies that cut their checks. Eventually, he does the Escalante country and in his hands the place becomes blood red with Triassic rock and is surrounded by a ribbon of Jurassic. His map hosts wonderful names—Waterpocket Fold is more brutally seen as Impracticable Ridges, the Kaiparowits is simply Linear Plateau.

What all these ambitious men scamper over is one of the planet's oldest and most stable layer cakes of largely sedimentary rock. They invent a language to record this place and the terms become the keys of the piano on which they play out their geological concertos: on one section of the Colorado, in what is now Grand Canyon National Park, the keys are granite, schists, Bass limestone, Hakatai shale, Shinumo quartzite, Dox formation, Tapeats sandstone, Bright Angel, Muav limestone, Temple Butte, Redwall limestone, Supai formation, Hermit shale, Coconino sandstone, Toroweap formation, Kaibab limestone, Moenkopi formation, Shinarump conglomerate, Chinle formation, Wingate formation, Kayenta formation, Navajo sandstone, Carmel formation, Entrada sandstone, Dakota sandstone, Tropic formation, Wahweap sandstone, Kaiparowits formation, Wasatch formation. In other areas there are more keys or fewer keys, but the basic structure remains the same.

The men stand back and look at the work and think they can hear the beginning of time itself, hear the thing ticking in the Precambrian rock too old to imagine and too hard and real to deny. They have named things, and now, they feel, they know them because of these names. This play often consumes their entire lifetime. Wheeler himself is broken physically by the work and retires in 1888.

They are the West we can truly possess. I am holding Wheeler's Escalante country in my hand at this moment. It is labeled Atlas Sheet 59 [*which reaches east to Capitol Reef*] and captures the dust and heat and thirst of the expeditions of 1872 and 1873 in fine lines, reds, greens, oranges, browns, yellows, and blues. The headwaters of the Paria move very exactly across it and plunge into Paria Canyon like an eighteenth-century minuet. The maps always humor us with their clean order.

But something happens in this hard rock country, even to professional military officers. Clarence Edward Dutton, Captain of Ordnance in the U.S. Army, loses himself in the surveys of the plateau country during the same years Wheeler is roaring about the slick rock. He produces volume after volume of reports, especially his *Tertiary History of the Grand Cañon District.* But science is not enough here. He writes:

> Whatsoever things he had learned to regard as beautiful or noble he would seldom or never see, and whatsoever he might see would appear to him as anything but beautiful and noble. Whatsoever might be bold and striking would at first seem only grotesque. The colors would be the very ones he had learned to shun as tawdry and bizarre.... But time would bring a gradual change. Some day he would suddenly become conscious that outlines which at first seemed harsh and trivial have grace and meaning; that forms which seemed grotesque are full of dignity; that magnitudes which had added enormity to coarseness have become replete with strength and majesty; that colors which had been esteemed unrefined, immodest, and glaring, are as expressive, tender, changeful, and capacious of effects as any others. Great innovations, whether in art or literature, in science or in nature, seldom take the world by storm. They must be understood before they can be estimated, and must be cultivated before they can be understood.

We inherit this split tradition of rocks we can name and things in the rocks we can neither comprehend nor ignore. We try poetry, music, photographs, paintings, drugs, alcohol, novels, tracts, guides. Nothing works. There are trilobites swishing in the muck, cycads are growing big as houses; it is all in the rocks, they have left their footprints and we cannot deny them their existence. This is what our brains tell us is embedded in this ancient stone. And more. Dinosaurs, yes dinosaurs also, roaring, making big imprints in the mud, dying and leaving large bones, frightening teeth, huge lizards that

Hiker in Sheets Wash, 1999.

make us shudder, all this is there in the rock. The initial humming is, well, we really don't know if it truly is, but we imagine it is some kind of humming right in the rock, in that oldest Precambrian stuff, the beginning of our world, the original hard rock cafe, and we reach out and touch it and the rock is hard and cold and yet somehow it scorches us. Then comes... well, everything, the entire catalogue of life, something vaster than Noah could imagine, something more diverse and richer than even his fine rainbow could convey. In the rock, you are trudging down a canyon, your hand moves, it touches and rubs...the rock. That is what our science tells us, that the book of life is right before our eyes, and recorded in what we determine is cold, unfeeling rock.

We all know this and yet none of us believe it. It is too immense, the units of time are too large, the imagination required exceeds our minds. But it seems to be so. We stand on a cliff and look out at the vista and we think this is beautiful, this is powerful, this is color and texture and a hot wind on our face, and cool water is running silently across the sands in the deep canyon below and we cannot deal with the ideas the rocks are whispering to us. We cannot, no, we really cannot. But here we come face to face with them and so no matter how long we go away, we always come back. If only in our dreams. Huge beasts, dead tens of millions of years, are caterwauling. A mountain bike zips past, the rider encased in modern fabrics. I'm warning you, don't come into this country for a vacation. It will take more. Ask the people rolling those wagons down toward the river to found that new settlement, the people with their children, their ducks, their chickens, their livestock, their musical instruments, their faith. It will take much more than you can give in a vacation. You will be back, I'm sorry, but you will be back. The beasts in the stone will beckon you, the humming of the deepest rock will wake you in the middle of the night no matter how far from this ground you roam. Fair warning has been given. It is not simply one more place. So we must come back for something we truly cannot name and we come back to a place we cannot really describe—even when we have memorized all the strata slowly humming with billions of years of energy under our feet.

17

## THE NEVERLASTING HILLS (1987)

MICHAEL COLLIER

*From his base in Flagstaff, writer, photographer, geologist, pilot, and physician Michael Collier has explored the West for forty years. This excerpt comes from his 1987 book for the Capitol Reef Natural History Association,* The Geology of Capitol Reef National Park. *"Some people think of science as sterile," Collier says. "Some people think of rocks as inanimate. But I can't write about or photograph a place without having some sense of relationship to it."*

Geologists are fond of leaning back, laying their hands across their bellies, and pronouncing that we can never really appreciate the full extent of geologic time. It stretches back too far to be truly understood. Nothing in our short lifetime can prepare us for the immensity that is measured in billions of years. There is a professional smugness here: the contemplation of time, we are being told, is the geologist's turf.

On the other end of the spectrum, faced with events that have occurred in a historic time, a geologist is likely to say that we can't really put them into perspective; we don't know if this flood or that rockfall was a fluke of nature or a true sign of the changing times. His feet are propped up on his desk now; he is nodding sagaciously and relighting his pipe. You see, it's all done, not with mirrors, but with the sliding scale of time.

On the 22nd of September, 1897, a flood raced down the lower Fremont River, submerging most orchards and fields that lay alongside its banks. The Mormons had met and risen above adversity before, but this was different. The rules had changed. Starting with that single cataclysmic flood, the Fremont began to entrench itself into a previously shallow meandering course. The small washes that feed it were suddenly spurred to slice open deep arroyos—one tributary to Sandy Creek, previously an insignificant wash, is now walled within a channel forty-five deep, seventy feet wide. The Mormons' small dams across the Fremont repeatedly silted up, and their irrigation ditches were choked with silt when they ran at all. Water supply, always a problem along the erratic Fremont, became undependable; the groundwater table dropped. Church records show that the 1893 population of the lower Fremont valley was 552 persons. Seven years later it had fallen to 372. The capricious land had driven out almost a third of its inhabitants; one hundred and fifty more would leave over the next twenty years.

With the flood of 1897, a cycle of erosion had begun along the drainage of the Fremont. The arroyos are apparently still cutting deeper into their beds. Tentacles of erosion creep further into the as-yet undissected flatland above the arroyos. The entire Southwest has been caught up in a regional version of this frenzied erosion since the 1880s. Is it due to overgrazing, or to a drop in the whole region's water table, or to climatic changes? No one knows. The lesson we learn is that the earth lives by its own cycles, its own rules. Processes that we observe now may have operated in different fashion or at different rates in the past. Cycles of erosion. Cycles of glaciation. Cycles of mountain building. The geologist's sense of perspective is nothing but a gambler's ability to play the odds. He knows that in the end the cycles will all average out. In only this sense is the present a key to the past....

*Erosion* on the other hand is the process whereby the products of weathering are carried away by wind or water. Grains of sand, particles of clay are airborne on every breeze, and are sent downstream with every passing rainstorm. Show me a storm-fed wash on the Colorado Plateau and I will show you muddy water. Wind is capable of reshuffling loose sand on the surface, but it is water which carries out the greatest work of erosion.

Hiking on a hot summer afternoon, it is admittedly difficult to imagine yet another agent of erosion: ice. Glaciers covered the top of Boulder and Thousand Lake Mountains in the not-so-distant past, perhaps as recently as 8,000 years ago. Striations and grooves on Boulder Mountain indicate that most of its ice flowed to the south, pushing rock and rubble down as low as 6,600 feet above sea level. A lobe of glacial outwash near Fruita left basalt boulders twelve feet in diameter littering the banks of the Fremont River.

Basalt boulders above Deep Creek, looking toward Thousand Lake Mountain, 2012.

Lakes were gouged into the top of Boulder Mountain, and were trapped behind the ice-inspired rockslides on its east flank. These too are the legacy of the glaciers.

Earlier, three characteristics that usually shape the results of erosion on the Plateau were listed: aridity, uplift, and flatness. We can now consider a fourth: the varying hardness of individual strata. The Mancos Shale is made up of alternate members of sandstone and shale. The shale is easily eroded, forming gray slopes. The sandstone resists erosion, forming cliffs that protect the underlying shales. As Dutton noted, erosion focuses its attack not on level ground, but against edges, against the cliffs and ledges.

By and large, the San Rafael Group is made up of soft shales and siltstone. It is more easily eroded than the underlying Glen Canyon Group with its resistant sandstone cliffs. As erosion strips down through this country, the Navajo and Wingate sandstones tend to linger above the landscape just a little bit longer than the other strata.

To the west of the Waterpocket Fold, small streams wander across the Moenkopi and Chinle formations as if confused, disoriented. When they reach the orange wall of the Wingate Sandstone, they pause a moment, thwarted. Then the Tapestry Wall at the head of Capitol Gorge funnels these streams into a purposeful plunge, headlong through the Fold toward the low deserts to the east. Many hundreds of thousands of years ago, erosion had not had a chance to cut down to this level. The Moenkopi and Chinle were still buried here; the wall of Wingate Sandstone did not stand above the countryside as it does now. Lazy streams meandered across the top of the Fold. But in time, the Chinle and Moenkopi were eroded relatively quickly; streams able to cut only narrow canyons in the Wingate and Navajo were able to erode out the broad valleys we see today in the Triassic shales on the west side of the Fold.

Deep, Sandy, and Halls Creeks all flow along the east side of the Waterpocket Fold, parallel to it, mainly through the soft layers of the San Rafael Group. If these creeks were to cut straight down from their present positions, they would soon be bedded in the harder Navajo Sandstone. But by a sleight of hand, they have so far managed to evade the Navajo. Instead of cutting straight down, the creeks find it more expedient to cut down and to the east, thus the stream beds shift east to precisely follow the Fold as it dives into the earth. At its Narrows, however, Halls Creek was caught off guard; its meanderings were trapped within the walls of the Navajo Sandstone. Gilbert describes the Narrows: "The traveler...can follow the course of the water and be repaid for the wetting of his feet by the strange beauty of the defile. For nearly three miles he will thread his way through a gorge walled in by smooth curved faces of the massive sandstone, so narrow and devious that it is gloomy for lack of sunlight." Such is the fate of streams and travelers in the Navajo.

The Navajo Sandstone, where it is tilted and exposed along the Waterpocket Fold, is intricately sculpted into a myriad of slot canyons, rounded domes, and free-standing arches. The walls of the canyons are sheer, rising hundreds of feet before rolling back to the domed terraces on top. The bottoms are softened by sand amidst the slickrock, carpeted here and there with maidenhair fern and columbines. These canyons are the gems of the Waterpocket Fold. It would be worse than pointless to direct you to the three or four "best" side canyons of the Fold. Nothing can replace the sense of discovery that follows one's own explorations. Go look, and you will find your own canyon within the Fold.

A friend and I once walked up Deep Creek, north of the Fremont River. We saw no one else; as usual in winter, Capitol Reef National Park was

Halls Creek side canyon, 2016.

empty. When it was time we turned up into the Fold and camped beneath a ledge at the top of the Navajo. Winter's chill was upon us; even so I couldn't resist jumping into a pocket of water just downstream from our camp. I should have known better.

The next morning we walked up into the heart of the Fold, the heart of the Navajo. A storm had cornered us under an overhanging arch. I would have played my flute to hear its echo on the surrounding cliffs. But I don't play very well and at my friend's insistence, had left the instrument back in the car. Instead we watched the rocks and listened to the rain.

Rivulets coursed across the slickrock, down paths well worn by rains before. We were within a wide amphitheater—the walls vertical, its floor by giant steps leading back to the narrow canyon that had afforded us passage half an hour before. Beyond the canyon we could see out to the Hartnet and South Desert where we'd been hiking two days earlier. There lay the San Rafael Swell and the gentle slopes of Moroni. An immense and solid country I thought, remembering the words of Gregory in the 1930s [*Herb Gregory, eminent Colorado Plateau geologist in the generation between Grove Karl Gilbert and Charles Hunt*]. But lowering my gaze to our amphitheater, I knew it also to be beautiful and delicate. It's all done with scale, both in time and distance. I thought of the rain trickling over the sandstone in front of me, and of the eons-long erosion of sedimentary rocks a mile thick. I looked at a small swirling convolution in the block of Navajo at my feet, formed by gravity before time had cemented its sand grains into sandstone—and then thought of the much larger Fold, thousands of feet high, ninety miles long.

Time and distance are the dimensions of geology. To study the geology of Capitol Reef is to consider the sum of an infinite number of small events. The science of geology is like a telescope—held one way, it makes its object seem larger; held the other way, the earth becomes vanishingly small. To dryly report that the Navajo Sandstone is 190 million years old, is to reduce reality to a hatful of numbers. But to examine its age grain by grain is to appreciate the true width and breadth of time. This perspective we gain through the eyepiece of geology. It will grace the life of anyone with the will to walk and the heart to listen.

The rainstorm never did amount to much. We walked back down to camp.

18

# A FIELD GUIDE TO BRAZEN HARLOTRY (2002)

ELLEN MELOY (1946–2004)

*Ellen Meloy's first book grew out of repeated trips through Desolation Canyon with her river ranger husband. Her subsequent books rippled outward from her home in Bluff. Meloy's thoughtful musings about Colorado Plateau natural history while she wanders the Navajo Sandstone of Comb Ridge in Bears Ears National Monument apply equally to the Navajo Sandstone spine of the Waterpocket Fold. This excerpt from the Pulitzer Prize–nominated* The Anthropology of Turquoise *introduces her wry humor, her scholarship, and her eloquence.*

For reasons that are not entirely clear I have always believed that love and restlessness are inextricably bound to a desert plant called cliffrose. The association might sound more plausible in temperate or semitropical places—all those sweaty, hothouse colors, the rising sap, the deep-throat orchids, roses with their beads of dew and libidinous landscape of lips and folds. In contrast, desert flora are sparse and ephemeral. There are spines, thorns, uncertain seeds, long periods of dormancy, and, when moisture comes, a passion of flowers so accelerated, you feel their demands on your heart, the mounting pleasure, the sweet exhaustion.

I first wallowed in botanical eroticism years ago, when the long, cold winters of my Montana home propelled me southward to cheat. By changing latitudes I could begin spring early. Beyond the northern Rockies' hard grip, the Colorado Plateau deserts were warm and soft, already engaged in rampant foreplay. The business of every plant, tree, and creature was the business of reproduction. In those years as a seasonal gypsy, on long hikes down canyons and across open mesas, I wandered through the fecund glory and tried not to eavesdrop. I committed Victorian acts of science: counting petals, defining shapes and symmetries, sorting the petiole from the pappus, the basal rosettes from the pinnately compounded. I indexed my brain with genera and spilled Latin names onto the Halgaito Shale.

The wildflowers taught me color, not lipstick or wall paint or inner-self color, but all the colors possible in a panorama of polychromatic reds held under a dome of azure sky—papery white prickly poppies, blue-violet flax, firecracker-red penstemon, the crimson spark of paintbrush amid silver-green sage.

The pages of my plant identification book wore thin from keying. My eyes lost focus in the dense undergrowth of footnotes. My sketchbook filled with portraits of rocks and plants. Next to the sketches of Anasazi pottery sherds and pinyon pine cones were notes like "Sinuous canyon, drowning in light. Navajo Sandstone? Entrada? Walls like flesh. The slickrock holds me in feverish thrall and the grip of inertia...'you who beneath his hands swell with abundance' (Rilke)." Or, on sojourns that coincided with a relationship in which I clung to a rather obsessive, unrequited love, "Datil yucca. *Yucca baccata.* Curled fibers on the margins of swordlike leaves. Center stalk with creamy blooms. Fibers used by the Anasazi for sandals and paintbrushes, roots used for shampoo. Liquid as slick as soap beneath a coronet of spikes. Extreme pleasure is a form of self-immolation, desire its leash around the neck. Do not bother to resist."

In no time scientific inquiry fell by the wayside. Phrases like "pubescence of the leaves subtomentose or tomentose" brought out a blush instead of a dictionary. I abandoned field guides, lists, and, near water, clothing. I spent a great deal of time lying down on dunes of coral red sand, shaping myself to the slip face and its crop of silky white and pink evening primrose. On one spring sojourn I mixed camp in a field of purple scorpionweed with a decision about marrying the man in my life, a very sound, un-unrequited love. There was little doubt in my mind what all these plants were up to, their wild, palpable surge of seduction best absorbed by the undermind—no categories, no labels, no conscious grasping but a kind of sideways knowing. Spring in the desert grew beyond the reach of intellect and became a

Desert paintbrush and cryptobiotic soil crust—the living shield woven by microorganisms that stabilizes, fertilizes, and prevents erosion, 2001.

blinding ache for intimacy, not unlike beauty, not unlike physical love. Even though I moved away from Montana and now live in the desert full-time, each spring still carries this craving, this weight of bliss.

And, here on the Colorado Plateau, once again the orb has slipped. The Northern Hemisphere tips away from the cold. The season is turning and, in my neighborhood, the lavish scent of damp sand, the upsurge of chlorophyll, the birds that mate in the air above our heads—all of it makes us slightly crazed. It is spring again. I have decided to live inside the sex organs of plants.

DESERT PAINTBRUSH (*Castilleja chromosa*):
A low-lying herbaceous plant with narrow leaves and clustered bracts, or modified leaves, that resemble a ragged brush dipped in scarlet paint. The bracts, not the flowers, bear the showy red. The flowers, as thin as eyelashes, lie hidden in the bracts.

At a distance the land clings to its wan winter brown, fired by the orange and maroon forests of leafless willows along the banks of the river below my house. Close up the colors come as fragments, as tendrils of green against red sand or onyx bibs on the throats of meadowlarks that change their guttural *tchuck* into the throaty melody of breeding season. The red-winged blackbirds, too, have traded winter songs for mating arias and flaming-red epaulets. Only days earlier their shoulder feathers were dark and drab. How do they grow the red?

...On a clear, calm day I hike up the tilted sandstone spine that knifes through the desert, the ridge of my winter quarters, where I brought moods and crayons and watched 360 degrees of broken desert in the ever changing light. A month has passed since my last visit. I missed the thaw, the shift from frozen water and dormant wood to a softening so full that the ridge itself could have swelled and doubled in size if I didn't know better.

On the ridge, spring's colors break out of the sandstone like a hulking seed bursts from a pod too weak to contain it any longer: aquamarine lichen, the fresh, sweet green of single-leaf ash, a scattering of cryptantha with blooms the color of the inside of a lemon skin, and, in sinuous folds of slickrock, wildflowers in electrified red: paintbrush.

The desert paintbrush is the queen of slickrotica. You can tell by its fiery scarlet and early bloom, as if it wants these curvaceous sweeps of sandstone to itself before the wildflower season's full Baroque. Paintbrush is partially parasitic on the roots of other plants. Underground, it invades the vascular tissue of another plant and absorbs its nutrients. Sometimes paintbrush nudges up seductively close to the host, a flashy scarlet starlet in pickpocket position.

In contrast to places where soil is abundant and deep—and to the neighboring Great Basin's bowl full of alluvium—the Colorado Plateau desert is earth unmasked. Weather and time remove detritus as small as sand grains and as large as hopes for a few inches of soil. On sedimentary strata with little or no veneer, plants like the paintbrush take purchase. They must adapt to different parent materials, fix their roots into substrate that varies in type and hospitality within a few hundred yards. Some plants take to sand and gravels, others to shallow silts or clay. "Edaphic endemism is rampant on the Plateau," says a statement about as tabloidesque as a botany text will dare. In other words, the range of certain endemics, or flora limited to specific localities, is often determined by soil conditions.

Paintbrush grows in tucks and folds of canyons and amid pinyons and junipers on the high-elevation mesas. A tall, gangly *Castilleja* with red-orange bracts prefers stream bottoms, springs, and other wet places. Paintbrush genera spread themselves from Wyoming to New Mexico and eastern California to Colorado. But many of them slip their lives into bare-boned sandstone. The paintbrush becomes attached to its homestead. I interpret this as affective as well as physical and take them on as allies. I admire their loyalty to dirt.

Red is common among early bloomers, as if nature wished to jump-start spring, to skip the formalities and lunge into an unfeigned passion that dissolves reason or reluctance. Red is the color of attention. Red flowers sear retinas made weary by winter, by snow or the season's low, angular light. The color prepares you for the wings of a bluebird, the shimmering violet throat of a hummingbird, sunrise in bands of pollen yellow. . . .

*Castilleja* produce nectar twice a day, when insects and birds are most active. Midday the pollinators rest. I am on the ridge at midday, greedily hoarding the red. The sun illuminates the paintbrush against blond rock. Red is the color of martyrs, blood, hell, and desire. It quickens the heart rate and releases adrenaline. The rains and this electric flower promise more red—and every other color—to come. I make a vow that each day will have in it a jewel like this one, knowing well that the desert bloom is not so much a season as a moment, given then lost. I am ready to live inside this paintbrush, but I cannot climb in. I am stuck in the ecstasy of anticipation....

CLIFFROSE (*Cowania stansburiana*): A small evergreen with gnarled trunk and shaggy limbs that carry five-lobed, resinous leaves. For most of the year it is unremarkable. (If you do notice it, you end up counting leaf lobes, one of the few features that distinguish cliffrose from its look-alike, bitterbrush, *Purshia tridentata*.) For a week in early summer, however, dense clusters of blossoms engulf every branch and twig until they fully conceal the leaves and the cliffrose looks like a pale yellow torch.

Cliffrose often prefers slickrock and shallow dry washes, where the embrace of low-slung rims on either side provides not so much shelter as a degree of difficulty, perhaps, to match the cracks and soil pockets in which they grow.

The first time I heard a Navajo say the word *gad* I thought it referred to a deity. Instead, it is the word for juniper, the omnipresent tree that grows atop mesas and in folds of wind-smooth sandstone across the Colorado Plateau. From juniper seeds Navajo mothers made their children bracelets of "ghost beads" to prevent bad dreams. The hard, fragrant berries bear one of the desert flora's scarce displays of turquoise. Fallen, they surround the juniper in a blue-green skirt. Here with *gad* grow companion pinyon pines, single-leaf ash, Mormon tea, and, on the slickrock ledges above a heart-cleft canyon, a scattering of cliffrose in bloom.

Creamy yellow flowers cover the branches from shaft to tip, drowning the small leaves in petals. The trunk's bark is brown with a silver cast, shredding in thin dry strips. Each flower has the broad, five-petal bowl typical of a wild rose and a crown of golden stamens in the center. I sit under a dazzling mass of roses in a breeze that is nearly a wind. It gusts, then subsides with rushes of sound like airborne surf.

Bees in the cliffrose fill the quiet parts of the gust rhythm. They are delirious and so am I. The cliffrose fragrance envelopes us in a spicy musk that is stronger, yet more delicate, than the water-heavy perfume of Russian olive.

It incites blatant acts of sensuality. Other plants prompt reactions that are aesthetic, intelligent, or herbal. Not cliffrose. It conspires with the sweep of slickrock on which it grows. It is nondescript when not in bloom, irresistible when it is. Sit by one and your heart will open and desire will flood into the emptiness created for it.

From this tree other cliffrose follow fissures in the rock in a somewhat orderly direction—the creases offer more moisture and soil than the acres of bare sandstone—but four or five more pale torches escape the line and erupt in different places, so there are cliffrose everywhere until the land drops off into the sheer space above a deep, deep canyon, and, below my high perch, meets the emerald-green crowns of a cottonwood bosque in the canyon bottom.

Like pale beige sand against turquoise sea, the symmetry of the desert is calming. Perhaps it is the spaces that are just right, the perfect amount of air and ground between bursts of creamy torches. Each cliffrose stands alone—or with a skirt of peppergrass and scorpionweed—and invites me to consider it the most beautiful, the fullest and most aromatic....

In the years of spring escapes from Montana to the desert, I so often found myself sorting out my heart. Slowly, the senses had a way of overtaking the mind so that the perfume of cliffrose could push me into a realm of sheer pleasure. The sandstone was filled with secret veins that only the flesh could find. *Press the stone and feel its flex*, I thought, take the shape of its harrowing embrace. The sun would rise east of my hipbones. Color struck like blows. The light ratcheted up its blaze and flashed through each nerve, a net of flame, a field of exquisite friction. I was seized by something that could never be wrestled down, a feeling not unlike sorrow and ecstasy compounded. Then came the sweet drift of longing, and I hung on to the earth's curve for dear life, breathing the comfort of stone during my ride.

The desert in spring held such vehement feelings that I would take them north like pollen stuck to my lips and the tips of my fingers and recall a kind of insatiable rapture—sun blazing on red walls, plaintive cries of pinyon jays, a slip of jade river, the brazen harlotry of cactus flowers, all of it alluring and distant. I would walk away from that wild, heady rose, put days and distractions and the distance of three states between it and me, then, pressed against memory, the scent would become as Keats said of his lover: "Everything that reminds me of her goes through me like a spear."

# 19

## VIGNETTES FROM THE "THE NEAR-SIGHTED NATURALIST"

ANN ZWINGER (1925–2014)

*Ann Zwinger moved with her family to Colorado in 1960. With the eye of an artist and the patience and attentiveness to detail of an art historian, she fell in love with the West. She began writing about the places she adopted as her home landscape, including* Run, River, Run: A Naturalist's Journey Down One of the Great Rivers of the West, *about the Green River; and* Wind in the Rock: The Canyonlands of Southeastern Utah. *Her natural history vignettes and elegant drawings slipped impeccable science into books for general readers. She cared deeply about place, family, teaching, story, art, community—and the future of us all.*

*Zwinger writes of aspen growing in clones, as they do—genetically identical shoots connected underground by their roots. Aspen clones reveal themselves on hillsides in fall, turning color at slightly different times, creating a mosaic of green and gold, each clone distinct in tone. The largest aspen clone of all grows near Capitol Reef at Fish Lake, a single clonal colony known as Pando, estimated to cover more than one hundred acres, weigh six million kilograms, and grow from 80,000-year-old-roots, a strong candidate for the earth's oldest and heaviest living organism.*

### Aspen (1991)

In the deepening afternoon, we cross the east flank of Boulder Mountain in Utah, between the towns of Escalante and Torrey. We ascend through aspen thickets, incessant trunks in small, close clones, filigreed with pewter branches. Around nine thousand feet the thickets coalesce into hundreds of acres of aspen, clone after clone, *Populus* populating the world. A lone aspen in the middle of a knoll, cloneless, with no competition for light or water, spreads as amply as an oak.

We come here in search of a grove of giant aspen. We find them standing among boulders of ruddy red and charcoal that stud the bronze-leafed floor of the clone. Younger trees are here but stand apart, distanced from these huge ones whose size and dignity demand space, require deference. They are not flighty Aphrodite trees, dancing and sparkling, given to frivolities. These are the Zeus of the aspen grove.

I can't even get my arms around the boles. Clearly they have well exceeded their life expectancy of two hundred years, although it's often difficult to tell, even from a core, given aspen's inconstancy of growth rings. In unglaciated sections of the country, as here, many clones may have endured many thousands of years. Spared the rigors of Pleistocene ice, they may well be direct descendants of Miocene- and Pliocene-epoch clones going back fifteen million years. The larger, fewer-toothed leaves of Colorado and Utah aspen more closely resemble fossil leaves than those of aspen farther north or in the northeast.

Aspen, Boulder Mountain, 2004.

I have no tape measure, so I remove my boot laces, tie them together, and find them not nearly long enough to go round. I settle for bending my notebook around the trunk and marking how many times it goes. One seventy-inch-circumference bole works out to a twenty-two-inch diameter; others have similar girth.

For the first eight feet the trunks are blackened, deeply furrowed and cracked; above, coal-black diamonds and lozenges pattern the bark like giant Tarot cards. Where branches fell they left heavily browed eyes like those of brooding, monster bulls. Monumental and isolated, the giants stand aloof, as if enduring so many winters and so many summers sets them apart, renders them able now to communicate only with others who have seen the same ancient comets, endured the same cruel freezes, withstood the same devastating droughts. They gaze somber-eyed into the distance, the superb survivors, shrouded in dreams.

## Wind in the Rock (1978)

In damper climates limestone disintegrates easily, but in dry regions it becomes one of the most resistant of rocks. Where the finer sediments have been removed by repeated flooding, ledges of more resistant limestone are revealed. Water has a tendency to shoot off these ledges or "knickpoints" during flood, its force scouring out a depression at the foot, which sometimes retains the water for a while.

One such charming pool lies just a quarter mile upcanyon, shaped like the winged samara of a boxelder. Gentling down to a soft bottom studded with rocks, the deepest part is maybe five feet. The convex side lies against a rock ledge, reflecting it in dark perfection. A chute cuts the ledge, and moisture, outlined with white as always, drizzles down the slope beneath, a film of moisture that remains and works away at the rock. The walls of the chute bind a narrow slot of blue sky.

In the shallows of the concave edge of the pool, the silt is as slippery as pure grease, giving way to a velvety softness in the water, blossoming up deliciously between my toes. About six dozen tadpoles, one-inch bodies with inch-and-a-quarter tails, snorkel along the bottom, bodies slightly canted, siphoning up desmids and diatoms—minute algae—while their tails undulate above them. They have fat, bulbous bodies, some already with rudimentary legs. Their bodies are mostly filled with intestines, of necessary length to digest vegetative food. When they become adults in July or August, they will be carnivorous with much smaller intestines, an adjustment of physiology accomplished in the transformation from tadpole to frog or toad.

As they bustle about feeding, they send up minute clouds of silt, so fine is the bottom material. Occasionally one rotates and reaches up to the surface, almost standing on its tail, smacking the air. They move in a rather dignified and leisurely quadrille. Against their clustered movement tiny water beetles dash frenetically through the water, disappear into the bottom silt, then pop out again, like minuscule Mexican jumping beans.

Standing ankle deep in the water, concentrating on the tiny beetles, I nearly jump out of my skin when a tadpole comes and nibbles on the side of my foot, a gentle but very unexpected tickle. I reach down to pick it up—they move so slowly that it's simple to scoop one into my palm. It wriggles there, its top the color of onyx shot with gold, underneath gilt and bronze, encased in a crystalline coating—lovely and soft to behold and to hold, a small Cellini creation. In the water, when the sun catches the whole fleet of them, they glint rosy gold. As I stand still, several, like miniature vacuum cleaners, clean off the thin layer of silt settled across my instep.

Up on the shelf above the pool hunches a small red-spotted toad. This silver-dollar-sized creature seems made of dull jade embossed with enameled dots the color of garnets. These toads are most likely to be found on rocks or in rock crevices, taking to water only when disturbed. During breeding season brown "nuptial pads" appear on the thumb and inner fingers of the male which helps them to adhere to the slippery body of the female during mating; eggs are laid singly or in strings on the bottom of these shallow basins. From April to September, during or after rains and in the evenings, their trilling adds to the quiet music along the pools and puddles of the canyons.

# 20

## FEATHERED CULTIVATORS (1981)

RONALD M. LANNER

*Forest biologist Ron Lanner taught for nearly thirty years at Utah State University. He has a special interest in the ecological and evolutionary effects of mutualisms of birds (especially jays) and pines. He says, "A surprisingly large area of the earth is covered by corvid-dispersed pines, in North America, Europe, and Asia; yet their story is little known even to the natural-history-obsessed public." His work for general readers aims to counter that lack of appreciation for this remarkable relationship, and this excerpt from* The Piñon Pine: A Natural and Cultural History *does just that for the raucous world of piñon jays at Capitol Reef.*

To the Shoshone it was known as Tookottsi. Early American ornithologists called it Maximilian's jay, after its discoverer of 1841, the Prince of Wied. Generations of Westerners knew it as the blue crow, but today it goes by the name piñon jay (*Gymnorhinus cyanocephalus*). By whatever name, it is not easily overlooked in the woodlands it calls home. It is a large crowlike jay, dull blue in color, with a short tail, a slender bill, a raucous voice, and gregarious habits. Unlike most jays, it is highly social; compact flocks of scores or even hundreds of restless birds may be seen making rapid rolling maneuvers within a hundred feet of the woodland canopy.

The clinical details of the piñon jay's life history are prosaic and straightforward. Pairs nest from late February to April in nests built in piñon, juniper, or ponderosa pine trees. Four to five eggs are usually laid. The nesting females incubate the eggs while their mates form a male flock that feeds

Piñon jay, 1974.

together, periodically bringing food to the females. Both parents feed the newly hatched, naked, slate-colored nestlings. In a couple of weeks the young assume a grayish color, including the bill, legs, and feet. After leaving the nest, they remain drably gray until the fall molt, when for the first time they assume a bluish hue. In a year or two, after breeding for the first time, they finally take on the full blue plumage of an adult bird. The piñon jay eats many kinds of insects, berries, and seeds, including piñon nuts. But the way it uses the latter is extraordinary and has helped to shape the evolutionary history of the pine that sustains it.

Late in summer—about the end of August—the seeds of the piñon become ripe. At this time, the cones are still green and pitchy and tightly closed, but the seeds within have well-developed tissues and are of high nutritive value. They are ready to be used. By now, the nesting season is over and nestlings have learned to feed independently. Yearlings have completed their postnuptial molt and, resplendent in their new blue plumage, have fully regained their ability to fly. Thus, the flock is ready for action.

This readiness manifests itself in drama as the flocks wheel and sweep through the woodlands, harvesting the fruit of the pine. Cones are pecked loose from their branches and carried to secure perches where the birds noisily hammer them open and pick them apart scale by scale. The seeds are removed intact, up to twenty being stored temporarily in the jay's elastic esophagus. These seeds are then carried off to the flock's traditional nesting area, as much as six miles away, where they are placed in the ground. Typically, the jay thrusts several seeds into the litter of dead needles and twigs that makes up the woodland floor. Some are deposited on the south sides of piñon trees, where deep snow will not accumulate and where snowmelt will come early in the spring. The caching of nuts continues through the fall, and as the cones dry and open, the jays are able to continue the harvest by picking seeds from intact cones still attached to the tree.

Their choice of seeds is neither random nor indiscriminate. Piñon seeds usually come in two colors: light tan and deep chocolate brown. Tan seeds in an overwhelming majority of cases turn out to be empty: they consist of a shell containing only the dried-up remnants of an aborted embryo, the common result of an ovule pollinated by a pollen grain from the same tree. In contrast, nearly all of the dark seeds are filled, and most of the filled ones have sound embryos embedded in plump white endosperm tissue.

Piñon jays know all about this: they are piñon seed experts of long standing. They ignore tan seeds, not even bothering to remove them from the cone. Instead they concentrate their harvesting efforts on dark-coated seeds. Here they do not rely on visual cues alone: the piñon harvest is serious business and cannot be left to chance. The foraging jay will remove a dark seed from the cone and "weigh" it in his bill. If it is one of the rare dark-coated empty seeds, it will be discarded. But even if its weight measures up, it may still undergo another test—it may be "clicked" in the bill by a rapid opening and closing movement of the mandibles. Presumably, it must make the right sound, or it too will be discarded. "Bill clicking" appears to tell the jay whether a filled seed contains a normal endosperm, or a diseased or resinous one. Thus, the piñon jay relies on its senses of sight, touch, and sound to reject seeds that would be useless as food.

The fall ends with the woodland litter concealing thousands of piñon seed caches in the areas where the jays will breed next spring. Soon courtship begins, and courting males, chased and beseeched by *kaw*-ing females, feed unearthed nuts to their soon-to-be mates. After egg laying, the incubating females subsist largely on uncovered pine nuts; after the hatch, the young nestlings are also fed the produce of the piñon. Thus is the reproductive cycle of the piñon jay entrained by the rhythm of the woodland.

The relationship of the piñon jay and the piñon pine is not one-sided; it is clearly a symbiotic one that benefits both parties. The piñon tree accommodates the jay in several ways. It not only produces very large seeds of high nutritive value; it spreads the cone-opening period over several months, making those seeds available throughout the fall. Thus, the bird receives a high-fat, high-protein food in large quantities that can be stored for use at a time when few other foods are available. The concentration of piñon trees in nearly pure stands allows the jay to accumulate a large volume of food in the woodland area near where it will breed. The lack of a seed wing, of course, helps keep the seed in its place on the upper surface of each cone scale until harvested. In addition, a thin membrane of cone-scale tissue (spermoderm) acts as a small "blanket," preventing the seed from falling out of the cone by mere act of gravity. And the tree further obliges by opening its scales widely to expose the seeds to view and by coloring its empty seeds tan, thus advertising to the jay that no effort need be expended on them. The advantages that come the tree's way are equally important. Because they are wingless, piñon seeds cannot be carried on the breeze to the areas of bare soil they need for proper germination and seedling establishment. But seeds harvested by piñon jays are carefully placed in prepared seedbeds, where some will elude the jays' efforts at recovery. These overlooked seeds have a high probability of surviving and becoming trees.

The connection between the piñon jay and the piñon pine is more than a bit of curious natural history. It is, in fact, basic to the ecology of both bird and tree and has provided the dynamic for the evolution of all the piñon pine species.

First, some ecology. Piñon seeds are big and clumsy things, incapable of graceful flight on the wind. If left to its own devices, a tree would eventually drop its seed to the ground beneath its own crown.... In order to germinate successfully under conditions that rapidly become arid, the seed must be buried in the moistened soil. How the seed gets there is of no concern to the tree; but that it gets there is clear necessity. This is where the piñon jay, and other members of the "seed-caching guild" come in. By burying seeds for future use as food and by neglecting to remove all of them by spring, the jays set the stage for seeds to germinate.

. . . Natural selection can operate simultaneously on jay and pine, modifying both species in the course of time. It seems likely that jays, by selective action over the millennia, crafted a form of tree adapted to growth in semi-arid conditions, whose large nutritious seeds attract the very birds needed to plant those seeds and perpetuate the trees' existence. In short, piñons were invented by jays. The jays still hold the patent—and continue to collect most of the royalties.

Cattle drive, Scenic Drive, 1975.

# PART V

## HOME

### Listening to the Desert

Mormon colonization reached Wayne County in the 1870s. Latter-day Saint Church leaders sent pioneers into the canyons with the challenge to make the desert "rejoice, and blossom as the rose." Those who were "called" by the Church did their best, moving into the Fremont Valley from the west, settling first in the fertile flats of Rabbit Valley (in Loa, Fremont, Lyman, and Bicknell). From this stable base, they seeded communities downstream in increasingly rugged country, through Torrey and Teasdale to the little oasis of Fruita at what is now Capitol Reef National Park headquarters.

Deep in the Fremont River canyon, Elijah Cutler Behunin built a house for his family that still stands. He intended to use the cabin as a base for grazing his sheep along the Fremont in summertime. Ruby Noyes Tippets, his granddaughter, pictures what that life was like during the one year the family lived here full-time, 1883–1884. She also tells us what happened to the tiny settlements downstream toward Caineville, all abandoned early on when the Fremont River refused to run in a controlled channel, and devastating summer flash floods intensified by uncontrolled grazing upstream washed away the pioneers' farms and fields.

The inviting flats where Pleasant Creek pauses before plunging through the Waterpocket Fold drew the first full-time non-Indian resident to the future park, Ephraim Hanks, in 1881. National Park Service historian Lenard Brown provides background for Hanks's arrival. A Hanks son and grandson remember stories from their family's homestead. The Hanks place, Floral Ranch, passed through a series of owners, leading to Billie and Levi Bullard in the late 1930s, who sold to Lurt Knee in 1939, who created the Sleeping Rainbow Guest Ranch on the site. Lurt lived at Pleasant Creek until his death in 1995. I had the chance to interview both Billie and Lurt, and their stories here come both from my interviews and from oral histories collected by historian Brad Frye.

Writer and conservationist Chip Ward brings the story of this foothold in the wilderness one generation closer. He and his wife, Linda, came to Pleasant Creek in the 1970s, leasing the Sleeping Rainbow Guest Ranch from Lurt and Alice Knee for several years. The Wards came from the urban East and made the startling landscape of Capitol Reef their home. In many details, Ward's description of daily chores at Sleeping Rainbow in the 1970s differs little from Tippets's descriptions of the Behunin family's days along the Fremont River in 1883. The guest ranch is now the Capitol Reef Field Station, operated by Utah Valley University, and UVU writing classes generate a steady stream of creative responses to the park landscape.

Wayne County still claims a handful of ranchers, even if they make their livings as teachers or government employees or builders and run only a handful of cattle on the side. Since the 1972 designation of the national park, all but one ranching operation with a grazing permit in Capitol Reef has sold to nonprofit organizations that have retired the permits. The romance of the range remains, along with the right to trail cattle through the park in perpetuity. In "The O-Bar Drive," Ray Conrad, Wayne County's preeminent cowboy poet, celebrates stock drives that took cattle and sheep through the park from the summer range of High Plateau meadows to winter range on the scrub lowlands below the Henry Mountains.

# 21

## WHITE HOUSE ON THE RIVER (1962)

### RUBY NOYES TIPPETS (1916–1997)

*Born in Torrey, Ruby Noyes Tippets loved researching and reimagining her family history. She describes the stories in her book as "legendary," but "based upon facts as they were given" to her. Her curiosity must have been infectious, as her son, Don Fowler, born in Torrey in 1936, became a distinguished historian and anthropologist.*

*Tippets's mother, Nettie Noyes, was the fifth child (of thirteen!) born to Elijah Cutler and Tabitha Jane Behunin (called Cutlar and Jane in her story here). In this narrative of her grandparents arriving at the bend in the Fremont River where they built their home in 1883, Tippets describes a visit to the Behunin cabin by Butch Cassidy. The outlaw who made Robbers Roost famous was born Robert Leroy Parker to a Mormon family in Circleville, ninety miles west of the park. For years, Cassidy rode through Wayne County to visit his family in Circleville on his occasional commutes from his desperado life east of Capitol Reef. Stories abound of his connections to Capitol Reef, including a log hideout in Grand Wash reputed to be an overnight refuge and the inscription described here by Tippets, carved on the back of the Behunin cabin.*

Jane was so tired that she felt as if every turn of the wagon wheels made an impression upon her back. It was not good, she told herself, to get this tired. She was forever reminding Cutlar and the children to keep the Word of Wisdom and here she sat so completely exhausted that she could not hold her head erect. Since baby Lila's birth she had never been well and she wondered with a feeling of lassitude, if she would ever be able to hurry as she once had done.

Cutlar had been excited about this move to a new claim up the river from Notom, and had promised her a large and beautiful home of white sand rock as soon as he could get it built. The older girls had looked searchingly into the tired eyes of the mother and she had sighed to recall the many times which her husband had made these rash promises. Cutlar had always hoped for a wonderful home for his big and hungry family, but each time he got one room finished, he seemed to put off building more for them.

The gray and cream colored ledges of this area were breathtaking in beauty and the river meandered along its quiet way through the valley. One could see almost any color he looked for in this remote and secluded place and the hues of a kaleidoscope were to be seen. Indigo, pure shades of purple, ochre, sienna, red, amethyst, and heliotrope blended into intricate forms of the sandrock and here in a place where the valley widened, Cutlar brought the wagons to an abrupt stop. Jane's head jerked upright and she quickly glanced at the sleeping baby to see if she were awakened by the jerking of the springseat.

The baby opened her eyes and yawned. "Well, Janie, here we are!" exclaimed Cutlar; "Here we will have the most beautiful home in all of Utah Territory. The rock lies waiting to be quarried and time wastes…let's get unloaded."

Jane lifted her baby from her bed box and carefully lowered herself from the wagon. As she tried to straighten her shoulders Jane knew that the kink between them would be there for the rest of her life!

The smaller boys ran about looking at the holes in the sides of the sandrock walls, and Hite informed them that the wind and rain had made those holes. Twelve-year-old Mort soon arrived with the stock and, as his big sisters made a bite of lunch for the hungry youngins, their mother lay down in the shade of a huge clump of rabbit brush nearby. The fussing baby soon caused her to arise to her feet and as Jane glanced at the sky she hoped that a storm would not come before she got a roof over her head.

As a home in the open was established that day, Hite and Ligie worked steadily upon a dugout which might be used as temporary sleeping quarters for the mother and her littlest children. As soon as lunch was fixed, Cutlar

Elijah Cutler Behunin family. Courtesy Capitol Reef Natural History Association.

called all of them to family prayer and thanked his Heavenly Father for a safe journey and reminded his Father that he needed health and strength to accomplish his work. Jane's knees were too tired to arise from this position and she sat there in the white sand and ate a bit of the food which her Red-heads [*her oldest daughters, May and Nettie*] had fixed for her. Her tiredness was so extreme that she felt that if she closed her eyes, she might die.

Before daybreak next morning, Cutlar had his family out of bed and ready to start work upon the big house. Jane felt about as tired as she had felt the night before, and the bed upon the sand had not done too much to rest her back.

While breakfast cooked, Cutlar and the three elder boys made a quick survey of the situation and decided where to start to quarry the rock. A sledge hammer and chisels were used to break the sandrock into squares and as Cutlar tapped the lines along the rock, the rock broke quite evenly and there were great piles of the broken rock by evening.

Cutlar had stepped off the area of the first room and, as the children carried the heavy rock to the site, Cutlar laid them in lines. The fireplace would be at the east end of the house and a door and window would face north. Later, Cutlar decided, he would add rooms to the west and south of this first room. One day, he promised he would have glass for the window and boards for the floor, but for this summer they would just have a dirt floor and hang a quilt up to the window.

Jane accomplished her washing duties on the bank of the river. Her old blackened tubs were filled with water, the fire kindled and her home-made

Behunin cabin, Fremont River canyon, 2017.

soap used to scrub the dirty and soiled clothing. As she had only one washboard, Nettie and May took turns using it and between turns they scrubbed the clothes upon the rocks in the river bed. The clothes were hung about on the large rocks to dry in the hot sun.

As the Behunin men worked frantically upon the white rock house and hoped that the rain would stay away, a sudden cloudburst hit the workers. Jane hurried into the dugout with Lila and Earl, and the rest of the children crawled under the box of the wagons. The warm rain fell steadily and as Jane glanced at her Red-heads, she was aghast to see them soaping their long red hair and washing it as the rain fell upon them. These girls, thought the amazed Jane, would take advantage of every drop of rainwater to get their beautiful hair set into a clean and fancy bob, as river water was very hard.

As the room was built up to the square, Jane admired her husband's workmanship. This was a well-built cabin, she stated, and it was as sturdy as could be. As she admired it, Cutlar reminded her that it was being built to last one hundred years. It was daubed with mud and would withstand the rains and winds of this rough country. Poles were secured for the roof beams and rough boards were placed over these poles.

At last the room was completed and Jane moved a few of her prized possessions into it. Her post bed almost filled one complete side of the room and the children had to sit outside on the sandy bank of the river to play. The family ate meals outside.

Cutlar and Jane were very proud of the rock house and there was room for them and the two smallest children to sleep inside. The dugout had been enlarged to accommodate the older boys and the girls made themselves a bed in the old wagon box. This was a life of luxury for the Behunin family…a roof over their heads, a fireplace to cook in, and a water supply right by the door. Jane's old dutch oven was used to make her lovely and tender bread and she cooked her tasty stews over the new fireplace fire. As fall arrived, however, the river dried up and Cutlar was forced to again haul barrels of water for his family's needs.

…One of the first visitors to the white house was Cutlar's old friend Butch Cassidy. He arrived early one morning and said that he was on the run from the posse which had trailed him for miles the day before. He had traveled from the [*San Rafael*] Swell, he said.

Butch Cassidy inscription on Behunin cabin, 1975. (The Park Service plastered over the signature in the 1990s, believing it inauthentic. Even so, when I worked as a ranger at Capitol Reef in 1975, an older man came in to the Visitor Center and told us young naturalists working the information desk that he had been with Butch the day he signed his name on the back of the cabin. Regrettably, we didn't get his name.)

Cutlar looked at the outlaw with mixed emotions, recalling the many kind things which this ill-guided man had done for him. There was no need to try to convert this man's thoughts to a law abiding pathway and Cutlar asked the outlaw to look over a part of his claim while Jane fixed them some breakfast.

Butch looked at the back wall of the white rock house and carefully started to carve his name in a rock above his head. Through the stillness of the dawn Cutlar stood watching the gang leader as he carefully carved 'Butch Cassid' in one rock and added the 'y' onto the rock which was next to it. This outlaw, thought Cutlar, failed to cover his trail well, and when and if the posse arrived there, this name in the sandrock would tip the officers off to almost the exact place where Butch was hiding!

Jane called them to breakfast and as Butch ate his share of cooked venison and dutch-oven bread, with molasses, Butch quietly told them that he likely would not see them again. He was leaving the country, he stated, and would not try to return. Cassidy talked to the Behunin family about the respect he held for them and in parting he sadly told Cutlar that throughout his life he had looked for a woman as good and sweet as Jane! If he ever found a woman, he stated, who could cook bread as Jane could, he would buy her the world and all that was in it!

As the rosy dawn filled the skies at the new place on the river, Cutlar foresaw a great future for himself and his faithful family. In his mind's eye he saw visions of fields of ripened grain, orchards which would never be plagued by frost. He foresaw pasture lands dotted with fat milk cows and hills which would provide ample feed for his flocks of sheep. Soon now, Cutlar promised himself, he would add new rooms to the rock house and add floors to them.

At a meeting, Cutlar outlined a vague plan to cut a channel for the river to flow into. And as he talked to the other settlers about, they enlarged upon his ideas and it was decided then that the Old Dirty Devil (The Fremont River) did not need to meander from one side of the valley to the other, it could be channeled into a straight course and give the settlers about 50 percent more land to till.

As the men worked upon the new river bed, they dreamed of a green valley akin to Dixie. The climate was perfect here, the growing season long and frost free, and the seed, which had been gleaned so carefully, would produce vast crops.

The women and girls fixed hot lunches and dinners for the working menfolk and helped the men carry the heavy rocks from the lands. These good womenfolk found themselves too tired out by nightfall to sit about the fires and listen to the men plan the bright future for them. These homemakers crept into beds made of straw filled ticks and thanked their God for that day and its blessings...the morrow and its needs they left up to the Lord...and the men....

The day finally arrived when the river waters were started down through the newly made riverbed. Men ran and jumped into the channel and swished the muddy water into their faces and the children stripped off all but a few bits of clothing and waded about in the smooth stream bed. There were no rocks now, to stub toes. Here was a wonderful place to play. The girls shyly sat on the edges of the stream and dangled their toes and tired feet in the cool water. Someone remarked that Mr. Hunt should be there with his violin; and soon he arrived and a party was held there on the sandy side of a new river which the settlers told themselves would start a new way of life for them. . . .

The lands about the old river bed were cleared of brush and rocks and the folk to the east end of the valley went onward with work to channel their river as the first ones had done. The summer soon passed and orchards and fields of grain grew everywhere. Some of the men fenced their claims with rock fences which marched over the tops of hills and down through the lush valley. Zion had never seen workers such as these!

The second summer soon arrived and the fields and orchards were planted and a song of harvest was in the stout hearts of the settlers. The river ran peacefully down through the wide valley and upon the cream-colored sand hills, dotted with black boulders, one saw men and boys, as well as women and girls, rolling the boulders into lines to form fences between the farms. The rabbit brush was grubbed out of its places and garden plots produced vegetables and berry brush instead. A harvest party was held at Aldredge, Utah [*Aldridge, three miles east of Capitol Reef, was abandoned in 1900*] and toil worn hands kept time to the fiddles as these pioneers danced about in the soft sand. Tired feet, often blistered and bleeding, tapped out a rhythm of thanksgiving to God for a harvest in the hills. Life looked very promising to Cutlar and to Jane.

Jane had gained in health and strength throughout the winter at Aldredge, and she worked as she had seldom worked before to keep her load of work all accomplished. A school was held for the children and the winter soon passed.

As spring arrived again the crops were prayerfully planted. Green sprouts were soon visible and a sigh of hope and thanksgiving again filled Jane's heart. The rains came, as the crops flourished, and at first the settlers were thankful for the rain. But soon, great flash storms hit the hills and torrents of water ran into the newly made river bed. Panic filled the hearts of the pioneers as the storm became more destructive and the river became a powerful force, which poured new streams of rain water into the new river bed and the floods came.

Cutlar and his neighbors worked to keep the water under control but soon it was apparent to them that the houses, as well as the crops, were to be lost. Cutlar had not heard of the words 'soil erosion,' as such, but he and others saw the error of rechanneling the Dirty Devil River and the great destruction which followed caused all of the settlers to leave this valley which had looked so very promising to them.

# 22

## PLEASANT CREEK HISTORY (1969)

Lenard E. Brown

*National Park Service historian Lenard Brown completed his historical survey of then–Capitol Reef National Monument in June 1969, six months after President Lyndon Johnson expanded the park sixfold.*

Ten miles south of the Capitol Reef Visitor Center is the Sleeping Rainbow Guest Ranch. Located on Pleasant Creek, it provides those who stay there a spectacular view of the towering cliffs that make up a part of Capitol Reef. Soon after passing the guest ranch, Pleasant Creek plunges through the reef and Waterpocket Fold to emerge on the east side and join the Fremont River. The first settler on the land occupied by the Sleeping Rainbow Guest Ranch was Ephraim K. Hanks and his family.

Ephraim Hanks has a prominent place in Mormon history. Born in 1826, he served as a sailor on the U.S.S. Columbus from 1842 to 1845. Soon after his return to his home in Ohio, he joined the Mormon Church. When Brigham Young called for volunteers to serve in the Mormon Battalion, Hanks joined and marched with the battalion from Missouri through the Southwest to San Diego. From there he returned to Salt Lake City on July 29, 1847.

From 1849 to 1856 he carried the mail from Salt Lake City to the settlements on the Missouri, making the trip more than 50 times. As the winter of 1856 approached, the Mormon leadership in Salt Lake realized that many of the handcart pioneers were caught in the high mountain passes by heavy snows. E.K. Hanks was one of those who labored valiantly [*in Wyoming*] to save them. From 1857 to 1860 he was active in various endeavors for the Church. In 1860 Hanks settled down to farming and for the next 17 years lived in the region between Salt Lake and Provo.

In 1877 President Brigham Young asked E.K. Hanks to leave his ranch in Parley's Park and go down to operate Lee's Ferry on the Colorado River. Hanks and his third wife, Thisbe Reed Hanks, were expecting the birth of a child, their seventh or eighth. On August 14, Ray Elijah Hanks was born; on the 29th Brigham Young, President of the Church, died. In early September the Hanks family started south for Lee's Ferry [*on the Colorado River between Glen Canyon and Grand Canyon*]. Upon the advice of John Taylor, soon to be elected church president, they wintered in Burrville [*fifty miles west of Capitol Reef*]. In the spring of 1878 Hanks learned that the Mormon Church had decided against taking over the Ferry as a church operation.

Ephraim and his family remained at Burrville for four years. He tried his hand at several different occupations with varying degrees of success, but the cold climate and the 7,000-foot altitude did not agree with Thisbe. The winter of 1881–82 was harsh and many of their cattle froze or starved to death. In May 1882, Ephraim and his sons, Walter and Alva, went south

Pleasant Creek ranch site, 2015.

and east to some land Ephraim had found on Pleasant Creek near Capitol Reef. They planted crops and returned to Burrville. During early August the family moved to this new home 11 miles southeast of Fruita on Pleasant Creek. The tenth child had arrived three months earlier....

In August 1893, Ephraim K. Hanks was made a patriarch of the Wayne Stake and he served as a counselor to Bishop Giles for the next three years. Hanks died on June 9, 1896, and was buried at Caineville, 18 miles from the ranch. His son Walter was serving as bishop there. Thisbe Reed Hanks lived seven more years and died on July 23, 1903. During the last few years she spent her winters at Caineville and her summers at Floral Ranch.

# 23

## TWO MARSHALLS (1948)

### Sidney A. Hanks (1875–1949) and Ephraim K. Hanks Jr. (1896–1984)

*As a Mormon patriarch of his time, Eph Hanks had four wives (though he never lived with the fourth) and twenty-six children. Polygamy was an official part of LDS teachings from 1852 to 1890, and about one-fourth of church members lived in families with more than one wife. As important as Hanks was, he had no biographer when in the 1940s his son Sidney began condensing the stories he had collected with the help of his nephew "E. Kay." This excerpt comes from their 1948 book,* Scouting for the Mormons on the Great Frontier, *published a year before Sidney Hanks's death. Their protective love for their family and their faith encouraged them to shave the rougher edges from their story.*

E.K. Hanks considered buying Lee's Ferry, but moved instead to a location some distance east of Burrville, in Wayne County, Utah. This was in a secluded spot in a box canyon, large enough for three or four families. The site was...beautiful and awe inspiring, with its gigantic walls of solid rock and its lovely, clear sparkling stream of water called Pleasant Creek, which emptied into the Fremont or "Dirty Devil" River farther on down the canyon. Here were a rich, productive soil and a warm climate, adapted for fruit raising, gardening, raising sugar cane—in fact, for producing almost everything they needed except grain, and flour was easily obtainable through the crops they did raise. Eph could envision the small valley populated with several good Latter-day Saint families, with schools and the auxiliary organizations so vital to the education of children. He purchased the most fertile picturesque spot in the valley, but his dream failed to materialize in part, as other families moved away, leaving the Hanks family alone. The establishment of a school was, of course, not possible and the children were forced to be away from home during the wintertime.

The family worked early and late, however, at all possible moments, improving and beautifying the new home, which was named Floral Ranch. More than two hundred fruit trees were set out on the bench, a good-looking frame house was erected on a rounding curve of the same bench, far enough back for flower beds and lawn in front. Well-built corrals and sheds were provided for the horses and cows. Bathed in sunshine, the splendid gardens, fields, and orchards added to the grace and dignity of the massive cliffs, which seemed to ensure strength and protection to the little valley. The complete happiness and contentment of the family was broken into only when the children had to leave home to attend school in neighboring settlements. For several years they spent the winters in Teasdale, and they were always glad to return to the ranch home in the spring.

Many people came to the ranch and they always found there a kind welcome. It was no wonder that place was sometimes called "Hanks' Hospitable Ranch!" One winter Ephraim played host to five or six men who, because they each had more than one wife, found it necessary to hide from the United States marshals. He enjoyed their company.

One cold stormy night, Ephraim's son, Walter, rode a bronco down from Teasdale over the slippery hills and frozen streams to tell them that the marshals were on their way to the Hanks ranch, and would be there by daylight.

Eph's guests hurriedly packed their belongings and some provisions, and left each finding a safe hiding place within an hour or two. Eph was there alone to welcome the marshals when they arrived....

Ephraim Hanks was always anxious to have his children understand the gospel, but he and Thisbe, his wife, had very little to say in their hearing either for or against plural marriage. The children knew their father had two other wives besides their mother, and they knew there were children belonging to these wives. Nothing, however, was ever taught Thisbe's children as to the doctrine of plural marriage.

About the time Wilford Woodruff, President of the Mormon Church, signed the Manifesto which declared the Church would obey the laws of the land and solemnize no more plural marriages, most of the offenders against the polygamy law had been to jail for from one to three years, and had paid from one to three hundred dollars in fines.

The understanding was that they were allowed to support their wives and young children, but were forbidden to have any further social relations with them. Even at that, there were children born here and there who marked the parents as breakers of the law. Such people were commonly called "Cohabs," or those who practice "unlawful cohabitation." Many of the young families that became entangled with the law fled . . . to Old Mexico, where they were welcomed with open arms. They bought lands and built up a fine commonwealth in that country.

For one small group of men this situation opened up new opportunities that were not likely to be neglected. For the United States marshals and deputy marshals, all this was like gathering grapes after the vintage is over. For every conviction they secured, they received a fee of fifty dollars.

During this uncertain and unhappy time, two deputy marshals drove a light, open-topped buggy and a rangy team of horses out toward the bottom lands; a man was working, bent over, part way up the steep slopes that led to the bench above where the ranch house stood. The man straightened at sight of the buggy and team approaching. He stepped instantly behind a pile of rocks, and then he reached one arm as high as he could and waved it in three wide sweeps toward the south.

A half-grown boy below seemed to catch the signal, for he immediately turned and ran off in the direction of the Wright ranch, farther up the Canyon—stopping only to close and carefully fasten shut a couple of gates that had been standing open, as he came to them, one after the other.

And a wagon that was standing, one hind wheel blocked, on the slope of the bench land above the road, as the two deputy marshals turned in to cross the bottom lands—must have suddenly run back against the corral fence, for now it was crossways of the road.

"Hey!" said Pete to the other marshal. "Did you see who pulled that block out?"

"What block?" demanded Mack. "The wagon there." Pete indicated it with a motion of the head. He was driving.

"Say! Dog-gone!" Mack sat up straight. "A minute ago it was up above the road—and now it's across it. We can't get by—"

Wayne County pioneer William Morrell at Eph Hanks's Floral Ranch, late 1800s. Courtesy Capitol Reef Natural History Association.

Pete jerked out the whip. "Just what I was a-telling you. We'll have to drive up around through Hanks' lawn. And Wright— Wright'll have time to make his getaway," said Mack disgustedly.

That evening the two deputy marshals came to supper at Floral Ranch. Eph had invited them. After supper, the three men walked out on the porch.

"Take this congress chair, Pete," said Eph. "Guess you fellows are a little tired tonight—"

Just inside the kitchen door a boy stood close against the wall listening. He had heard that anyone shielding a "Cohab" was subject under the law to one year's imprisonment and a heavy fine. He did not like the looks of those two deputy marshals and his heart thumped when he realized how close he was standing to the two dangerous men.

"Mack and I," Eph went on, "will each take a sack of husks and sit on the bench."

There was the sound of chair legs on the floor. The marshals took out their bags of Durham and were soon enveloped in a cloud of smoke.

It might have been the spell of the fragrant peach orchard bathed in the gentle light of the moon, or the twitter of night birds calling their mates, or the faint sound of running water where Pleasant Creek glided through Floral Valley. It may have been the glorious sight of the full moon rising over the towering, two-thousand-foot cliffs that stood boldly across the valley. Or it could have been the sweet, young voices of the girls of the family, as they washed the supper dishes and sang their Sunday School hymn, "We are the bees of Deseret."

Something, at any rate, seemed to shut out the turmoil of the day from the hearts of the two visitors on the porch. Momentarily they forgot the prize that had barely escaped them this afternoon—forgot even that the news of their presence here was being carried hastily from ranch to ranch, warning all other polygamous friends to flee from the wrath at their door.

Breathing deep of the fresh spring air, Mack remarked, "Eph, you have a lovely little ranch down here."

"Yes," said pessimistic Pete, "if his little haven was only rid of these law-breaking neighbors."

"I'm afraid," Eph put in quickly, "you haven't the respect for our neighbor, Mr. Wright, that I and my family have learned to have."

"No," Pete snapped, "we haven't. I suspected you didn't want to see him caught when we found that wagon blocking the road."

"Wagon?" Eph inquired gently. "Yes," went on Pete. "The wagon across the road that forced us to turn up to your house, across your lawn, and back over your wife's rose garden. And your pressing invitation to stop with you while our horses rested, delayed us."

"Hold on a minute," Mack broke in.

"Don't accuse our host of aiding and abetting Wright's get-away."

Just inside the kitchen door, the boy's fingers gripped the doorframe and his ruddy cheeks paled a trifle.

"Oh no," said Pete, and he laughed quietly. "If it wasn't for our friend, Eph, here, we might be sleeping tonight out under some cottonwood tree, with no food for ourselves or our horses."

Mack cradled his pipe in one hand.

"You can't trace anything to Eph," he said.

"Sure not," admitted Pete. In a minute he added, "But them two gates we had to stop and open—no doubt they've always been kept shut." There was a faint sneer in his voice. "And that pair of cowboy boots, laying at the side of the road as we went up to Wright's house—well, of course they didn't show the owner could run faster without 'em on."

Mack laughed. "When we come back," he said, "those boots had got tired of waiting and had got up and walked off."

"Yeah!" Pete flashed a bitter look at Eph. "I wonder."

The boy inside the kitchen door winced and his hair may have risen a little higher on his head, as he glanced down at his own footwear and wondered if this man was a detective like the ones you read about in books and newspapers.

The boy muttered softly to himself, but no one could hear him. "Behind that pile of rocks Johnny Giles gave me three waves of his arm, and I guess there was nothing wrong about me running a half mile, and slamming two gates, and giving the three waves to Wright—sure there was nothing wrong—" He knew that Wright had ducked into the sagebrush immediately on the signal, but had later raised up incautiously and had then been obliged to make for the creek under cover of the sage. Sliding down the bank was quick business and, once in the water, the marshals, following his track, could not tell which way he had gone....

...Sidney, the boy who had closed two gates and had passed on the three waves to Neighbor Wright, soon lay sleepless where he could hear all that was said on the porch.

"Eph," said Mack, "you've got three families, but you haven't got three wives now. Do you mind telling us how you fixed this all up?"

"Well," returned Hanks slowly, "you fellows know us Mormons well enough to realize what a blow this new law against plural marriages was. Look at me, for instance. Some years ago I married Harriet Little, a young widow. I would raise children to Little—they would be his in eternity, as the Bible says, '...Moses said, If a man die, having no children, his brother shall marry his wife, and raise up seed unto his brother.'"

Mack nodded his head.

"Then," continued Eph, "under the law of our Church, I later on married a young girl, Jane Capener. Her children will be mine through all eternity, as she will be my wife, worlds without end. Harriet fully understood this and gave her blessing to this union. A few years after this, I married Thisbe

Reed, who was one of the members of the Handcart Company, coming over the mountains. My two wives gave their consent and adopted Thisbe into the family. She and Jane, with their children, are sealed to me for all eternity."

The cry of a night bird wavered through the stillness and in the front room, wide awake on his bed on the floor, his brother asleep beside him, the boy Sidney thought what a busy man his father must have been in those days—carrying the mail and fighting Indians, while he had three homes and three families to take care of!

"But," said Pete thoughtfully, "you haven't got all three wives now." He straightened suddenly in the congress chair. "Or have you?"

"No," replied Eph. "But those three wives have got some mighty fine sons. There is Harriet, for instance, with George, Sell, Pep, and Charley; and the oldest are grown men and able to do for their mother. Jane has Bill, George, Eph, and Dave. Thisbe's oldest son is Walter, and he and I have made a pretty fine ranch out of a patch of sagebrush. I have yet, even now Walter is married, plenty of help with Sidney, twelve, and Ray, nine, and Art still coming on for my old age."

"Well, I'll be darned!" Pete broke out, dropping the front legs of the congress chair to the floor with a bang. "You've got it all planned out—"

"Besides," Eph went on, "each mother gets help inside the house from her girls. You see we Mormons have a great advantage with our well-trained big families. The older ones unselfishly help the little ones, and we don't have anything to do but work and be happy—"

"But," put in Mack, "how about education?"

"Well, we manage pretty well. Walter's wife is a school teacher and she keeps our children up to their grades."

"Yeah, but," said Pete again. "You haven't got all three wives now—"

Mack moved uneasily. "Why don't you teach your children that your prophets can be wrong?"

"Mack Armstrong," said Eph, "I had rather have my arm cut off right there—" putting his other hand on his arm just below the shoulder—"than to teach my children false doctrine—"

"But hold on, Hanks!" Mack sprang to his feet. "Your Articles of Faith by Joseph Smith, your Prophet, say that you believe in 'obeying, honoring, and sustaining the law'—"

"I do. And that is why, when the Edmund-Tucker law [*banning polygamy*] became effective and was declared by the Supreme Court to be constitutional, I called my three wives together and laid the situation before them. I have fought for my Country and the Constitution," Eph went on quietly, "and I do not want my children to lose faith in that great document."

In the front room, the little boy, Sidney, grew suddenly sleepy. He cuddled down beside his slumbering brother and drew a long, tremulous breath. His heart was full of a quickened love for his father and a deep gratitude for his father's honor, which chose to sustain the Government of his Country, as the Articles of Faith of his Church demanded.

"What did your wives do?" asked Mack.

"My two first wives chose to get divorces which the court granted them free, and I re-married Thisbe."

"So you're in the clear! We can't touch you," grumbled Pete. "Too bad."

"One of them lives in Salt Lake, don't she?" Mack broke in.

"Yes, Harriet and her five children. And Jane in Heber City with her five. Then Thisbe and I came here and started Floral Ranch."

"And I'll bet you gave them—them others—support and comforts," Mack said.

"Well, a man naturally does what he can." Eph was looking off, far up the canyon.

Pete stood up beside Mack. He glanced around at the pleasant five-room house, which was the largest and finest house in Southeast Wayne County at that time. It was a frame house, lined with adobe and plaster—warm in winter and cool in summer. All the rustic work was made with a hand plane. The view from the wide porch was magnificent.

Pete grinned and nudged Mack.

"And we," he said, "have a hard time keeping one wife and two kids apiece!"

Sidney felt very happy and very secure. He was no longer afraid of the two marshals—or of anybody!

# 24

## TOO LONESOME DOWN THERE (1991/1994)

### Billie Bullard (1913–2005)

*Billie and Levi Bullard lived at Floral Ranch in the late 1930s. They sold their ranch to Lurt Knee in 1939. When I interviewed Billie in Bicknell in 1994, she told me she was raised in Toppenish, Washington (where an outlying community of Mormons worked in the sugar beet industry—and where my own non-Mormon father grew up). Her "folks had ten girls and sent them down to Utah to find Mormon husbands." She found one, Levi, on her second trip and they stayed married until his death fifty years later.*

*Like many rural Mormons, the families in Fruita and Pleasant Creek didn't necessarily follow the LDS "Word of Wisdom" that prohibited drinking alcohol. Stories abound of stills and beer- and wine-making. (From interviews by Bradford Frye, 1991, and Stephen Trimble, 1994.)*

We didn't have very much to live on. But the Depression was the best years of our life. No crime. Everybody was for everybody else. Nobody went hungry. Levi herded sheep part of the time. That was one reason we went to Floral Ranch. We thought we'd be together. But then I had to be at Fruita for school. So Levi was alone all week anyway.

In '37 my first son Francis was in the first grade. Anne Snow was a very good teacher in Teasdale and she supplied me with books, and I taught him the rest of the year. I didn't think I was doing well enough teaching so we decided we'd better take them into school. So in '38 we took them into school in Fruita and stayed in Doc Inglesby's cabins [*Inglesby ran a Fruita rock*

Ranger Diane Gifford at the Fruita Schoolhouse, dressed as a 1930s teacher for Harvest Homecoming, an annual celebration of Mormon pioneer heritage in the park, 1994.

*shop and motel*]. Jennie, the next one, there's only fifteen months between 'em, she was ready for school too, then. But being the littlest one in the classes she didn't get much attention. She was put in the corner painting with crayons. But it helped her 'cause she can really paint now. She sells her paintings.

The schools up above [*in Bicknell and Loa, in western Wayne County*], when books got old they'd send 'em to Fruita. So their textbooks were mostly from the schools up the county. The kids didn't have much homework when they'd come home 'cause I'd read to them a lot at nights. What we'd do—I have that *Geographic* magazine and with the children we'd go on trips. We'd get that out and imagine we was them people there....

When we bought the ranch, we'd all get together in Fruita every week or two and have parties. Ice cream parties and play cards. Then in summer, they'd all come out to the ranch and we'd have rodeos. We'd get our stock in and all the women and all would try to ride. I remember once we rode the calves and we'd all rode, and this one lady, Laura Mott, she was the last one to ride. This one calf was too tired to buck so they just told the dog to nip his heels and she just flew through the air.

I know they took a picture of me riding one of the steers. When the picture come back there was just a shadow in the air. I wasn't on the steer or on the ground, either one. Then we'd mostly have lots of food and have a potluck and a picnic. Everybody would bring food.

And then as you go out of Fruita, 'fore you get to Danish Hill, they'd had a big ball diamond. And we'd choose sides and play baseball with the older kids and the adults, both men and women. We really had some baseball games.

The guys made their own malt beer in kegs for the 24th [*Utah Pioneer Day, July 24*]. Levi tried aging his beer in my oven. It blowed up and the oven door blowed open.

This boy Johnny Larick, son of a good friend and foreman at the Bingham Canyon mine [*the huge open-pit copper mine in the Salt Lake Valley*], he stayed with us summers lots, and he and I would drive the cows up in the spring to feed up above in there some, and they'd drift back. And that ole cave was up in there. They told me, I don't know, that the polygamists used to go up in there and hide, in the polygamist days. And I rode up and back in again' the ledges one day, and tied the horse up and walked quite a ways. And when I was up in there I kept hearing something.

I'd look back, but I never did see nothing. But the next day or two, I was telling 'em about the country back up in there, how beautiful it was. So, Dick Wilson was staying in to Fruita and Charlie Gibbons—an oldtimer from Hanksville who stayed with Doc Inglesby and took care of him—so they came out and went back up in there. I don't know if they was kidding me or not. They swear it was the truth. They said, "Well, there was a cougar following your tracks down to the house. We could see it just as plain in places." So I decided that's what I heard.

I shouldn't tell you this, but we poached deer 'cause we didn't have meat. We's saving our beef and pork to sell to pay for the ranch. So that was our meat. And asparagus in the spring. We couldn't afford shells to shoot 'em, so you'd have a bear trap, and you'd wrap sacks around, you know, so it wouldn't be just the metal and we got two or three deer that way. We'd put 'em where they would trail, where they come into the field. But we never wasted it. We'd use every bit of it.

You could just raise everything down there. We raised squash down under the hill there from the house down in them few little trees. This Hubbard squash, one of them was 97 pounds. Doc Inglesby had a time loading it into the back of his car. He took it up to Salt Lake to the fair and got a first prize on it. Told 'em he raised it. He cut that squash up and gave to his relatives, pieces of it. They said it was really good.

You could raise melons and I raised a few peanuts and peas. Mostly, we raised pigs. We practically paid for the place in pigs. The pigs were running out and they couldn't get in the garden, but those little ones did. Their little snoots. There was just a row down them rows of where they had smelled them and got my peanuts and peas.

We raised lots of corn for feed. But it was so hard with the corn. The porcupines were thick. And they would come in and were just like hogs when they would get into the corn. This old dog…Levi would go out with him and he had a certain club, he had, and that old dog would go up and down the rows and when he found a porcupine he'd just start barking and then Levi—there's this certain place you'd hit 'em. On the head or somewhere, I don't know. But the old dog wouldn't bother them until Levi had hit 'em. But then Levi would have to get ahold of him or he'd get down there and get porcupine quills in his nose.

Oh, we had that old car and we'd come to Torrey. It was an old Ford. Model T, I think. I know that every time we'd come out why we'd have to stop two or three times for Jennie, she'd get so sick to her stomach, riding. So we didn't come very often. Then we could send with the mail carrier and get something if we needed it real bad.

I tell you one time, our own mules got loose just after we'd moved down. They went clear to Torrey. And Levi was riding the haying or something. He needed 'em, but he had to be home. So I went and rode my horse into Fruita, to Cass Mulford's. I caught the mail truck there and rode up to Torrey with him, took my saddle. Then I rode one mule and led the other back down and picked up my own horse and went back to the ranch.

I'd ride my horse when I'd go down to Notom, but I wouldn't just go down and ride back up the creek. It would be when I'd be going after the

Pleasant Creek gorge, 2008.

mail. I'd go over and go down to Notom and come back up to Capitol Gorge to where our road took off to the mail box.

Every other day the mail would come; Monday, Wednesdays, and Fridays. And I would ride my horse out to get the mail. I knew about when they'd come. When I'd go I'd just put my legs close to his flanks and boy he'd take off on a run. I'd go around those turns, he'd just lean, going around the turns. Dewey and Nell Gifford was up there one day visiting. They had to go back and I was going out after the mail. He left in his pickup and I got my saddle on my horse and passed him part way out. He couldn't believe it.

I remember once we had this corn patch. Levi had a cultivator that the one mule pulled but then you'd have to hoe between the corn. In them days you didn't see women wearing shorts very much. But it was so hot, I put shorts on, and was hoeing corn over there. And somebody came and I had the darndest time getting back around and coming up the back door to the house to get in so they wouldn't see me in shorts.

We left the ranch in 1940 because we got it paid for. Levi had to live alone all week long when I went into school, and we just didn't get married to live alone. So we moved up the county and bought another ranch, Fish Creek Ranch, between Grover and Teasdale. It was a beautiful place.

[*Floral Ranch changed hands often because it was*] probably too lonesome down there. It just makes you feel bad when you go down there and see those fields all like they are now. The alfalfa all's gone. The trees are gone that were down there. It just don't look like it did at all. And all of that big piece of ground underneath the hill from Lurt's. That was a beautiful field of alfalfa, new alfalfa, when we left.

# 25

## EATING THE VIEW (1991/1994)

### Lurt Knee (1909–1995)

*Lurt Knee ran his Sleeping Rainbow Guest Ranch at the site of Floral Ranch in Pleasant Creek for some thirty-five years and lived there for another twenty years. His favorite visitor was Josef Muench, the widely published landscape photographer who came each year and depended on Lurt's backcountry expertise to take him to underphotographed gems. Lurt's sister introduced him to the idea of guiding tourists as a profession, since her family ran Goulding's Trading Post in Monument Valley on the Navajo Reservation. Lurt's second wife, Alice, whom he married in 1958, was his equal partner at Sleeping Rainbow throughout the subsequent decades. (From interviews by Bradford Frye, 1991, and Stephen Trimble, 1994.)*

I told Harry [*Lurt's brother-in-law Harry Goulding, who ran a tourist operation in Monument Valley*], I said, "I've been reading, Harry, and I wonder if you've ever heard of this huge Waterpocket Fold that starts at Thousand Lake Mountain and runs through the country for 150 miles and becomes lost at Hoskininni Mesa?"

He said, "Yes, I've read about that."

I said, "Well, I read in the Park Service magazine that they're talking about making a national monument over there." It was made a monument in '37, and I got here in '39. I said, "Well Harry, I would like to stay here with you, but I want to find my own place that is near to a big red ledge as you are over here." I said, "I don't know just where. I want to get as close as I can to this national monument, I'll grow with it."

So he gave me a kick in the pants and a pat on the back and said, "Go to it." So I came over here and I didn't know then, but I came in here through South Draw down the old Anasazi Trail over the Miners' Mountain.

The uranium people [*in the 1950s boom*] built the road where it is today, but over this way a ways on the Miner there is a real rugged old abandoned trail that I came down over. I never knew that there was a ranch here, I just thought, well the water going down there, there might be something.

I ran onto this old ranch that Levi and Billie Bullard owned at that time. We stopped here and Levi said, "If you stay overnight we'll have a great feed out of the garden, it's just noon now." So we did that, and then he said, "You people look like you travel. If you know anybody that—"

"I didn't ask him first, I looked around and I liked the ranch, and sure enough we were adjacent to this national monument where its line was my line."

He said, "My kids are getting school age and I would like to sell it if I can."

I said, "Okay."

He said, "Would you live here where I'm living?"

I said, "No." At that time that old building was so terribly rundown. Skunks came out and ran all up over the dining room table every night. One night he threw his shoe at it to scare it.

Levi said, "I made a mistake, Lurt," he said, "I hit the damn thing." Then he said, "I had to keep my head covered up the rest of the night."

So he was that kind of a guy, and he said, "Where would you build?"

I said, "Right up on that hill."

He said, "Well, I wouldn't live in this old house myself any longer. I was going to build it up under those trees where the water will run to me. What do you want to build up there for? Why don't you build down here where the water will run to you?"

Which made sense of course if I was only looking at grass, a cow, a pig, and so forth. But I had other visions of what it would be.

I said, "You come back in a few years and I'll show you what I mean."

So he came back and he said, "You did build up here!"

I said, "Yes. Come here Levi. Look at that window there, that frames a picture out over Pleasant Creek in the valley where old Ephraim Hanks first started 135 years ago."

Lurt Knee at eighty-five, in 1994, a year before he died.

He said, "I've got news for you."

I said, "What's that?"

He says, "Don't you know you can't eat a view."

I said, "Levi, you come back in ten years and I'll prove to you that I can."

He said, "Now, I know that you are nuts."

So help me, he came back in ten years, and I guess it was Easter or something and I had cars all over this hill, and he came up through and worked his way around and got through and got over and he said, "Lurt, I had no idea that anybody would want to look at these old red ledges."

I paid $3500 for the Hanks ranch, and then I looked at the surveys, found plots for sale, and bought them at 15, 25, 50 cents max per acre. I preserved this land for the park.

The old house was falling down, the fireplace had fell over and the old walls would wiggle back and forth. It was sod, faced up with shiplap, and the old house was really rickety and dangerous. It could have been preserved, I think, but I was out in the winter [*of 1945*] getting all my different things together, giving my slide show in Arizona and California, and my help that I had here burned it down. I came back next year and he said, "Well, I got rid of the old house."

I said, "What?"

He said, "Yes, I burned it down because it was very dangerous."

Harry Goulding pulled "Sleeping Rainbow" out of the Navajo and gave me the name. The Navajo named it, but the Chinle, the most colorful formation in the earth crust geologically speaking—that's where we get it from.

Building up the place was just my job. I never hurt anyone in the backcountry in 50 years of trips. And I was always able to come in under my own power; I'm a shade tree mechanic and could fix anything that happened.

I'm quite an artist at making up an answer. I had to be! And I had to stay neutral. I never knew what religion I had in my car.

I always called it the "uranium scare" [*the 1950s boom*] rather than "uranium fever." I had almost 100 uranium claims between here and the Colorado River. At one time you might say there was a prospector behind every bush. Hundreds of them came in from all different areas. They were absolutely scattered all over this country. One came in here one day and said, "Are you Lurt Knee?"

I said, "Yes."

He said, "Boy I've hit some of the richest ore you could imagine, it's all over this country."

I said, "Brother, I don't think you know what's going on."

He said, "What do you mean?"

I said, "Well, out here on those old black boulders right now with your geiger counter it will read full scale."

He said, "Yes, that's what I noticed. This country is loaded!"

"No," I said, "you're wrong. We've had fallout from atomic blasts, and that's when they were testing over in—what's that island that they blew up?

Eniwetok. Anyway, they moved all the people off and set off some test bombs there, and we had a fallout."

This was the same time they were testing in Nevada at Frenchman's Flat. The two together; we had quite a fallout. He was very disappointed and he said, "Well, I didn't know that."

I said, "No, you've been out here prospecting and you haven't been listening to your radio." So it only lasted about three days, and the first rain came along and washed it away.

# 26

## LANDING UNDER A SLEEPING RAINBOW (2000)

CHIP WARD

*Chip Ward made his first landing in the West at Pleasant Creek in the 1970s. In subsequent years he drove a bookmobile, led campaigns to make polluters accountable, cofounded HEAL Utah, became the assistant director of the Salt Lake City Library System, and then retired to Torrey, where his western journey had begun. He's written a rollicking novel based on the cast of characters living today in a town like Torrey,* Stony Mesa Sagas. *His book* Canaries on the Rim: Living Downwind in the West *opens with this essay.*

A child of the East, I had never seen a desert. My eyes had been conditioned to see beauty in the pastoral landscapes of Vermont, the rich forest green of the Adirondacks, and the classic alpine vistas of Switzerland. As I approached Capitol Reef across an open horizon of gray and treeless mesas, I was struck by the bare, unfinished, almost ruined qualities of the land. "My God," I thought, "this is the world's largest construction site."

It was hot and I remember how Factory Butte and Caineville Mesa stood out as if they were all that was left of the surface of the planet after it was excavated. As we drove along, flat cracked and eroded ground alternated with huge graceful mounds of muted ash, elephantine evidence of ancient volcanic activity. Maybe it was the Moon. I turned to my wife, Linda, and said, "What the Hell was Bill thinking?"

We were on our way to Capitol Reef National Park to visit her brother

Bill who had just landed a job as a seasonal ranger. Eventually, desert gave way to sandstone canyons. We began to relax and look up. My worries about the mental condition of my brother-in-law subsided. I liked what I saw. Redrock is an acquired taste even if you enjoy your first bite. A couple of years later as a guest ranch operator, I had the opportunity to observe how first time visitors respond to Southwest landscapes that differ radically from the Walden Pond paradigm of Easterners. Most were intrigued and fascinated. Maybe their expectations had already been primed by those coffee-table books by photographers who captured the fractured and stained canyon walls as art. Maybe they had seen John Wayne riding across Monument Valley in "She Wore a Yellow Ribbon," or Robert Redford jumping off a slickrock ledge in "Butch Cassidy and the Sundance Kid," or any number of car commercials. Even so, it took them awhile to fully appreciate what was right in front of their eyes.

A canyon wall may seem beautiful on first viewing but it is only the first act. Canyons capture light and texture over time. The picture shifts from morning to night, with each passing cloud, and from season to season. Rain changes everything and so does snow. Rain creates dramatic and unexpected waterfalls, washes away dust to reveal new tones of color. Snow outlines canyon wall fracture patterns in bright contrast. Rocks also carry aromas that tell you if they are hot, cold, or wet. And they carry sound, absorbed or echoed, so that the voice of each canyon is different from another. Loving a canyon completely requires the time and patience to wait for all her moods and nuances to appear and reveal themselves to all the senses.

Capitol Reef's main claim to fame is its colorful geology, especially the multicolored layers of rock along the Scenic Drive. Spires and domes of white Navajo sandstone sit on top of big broken red walls. Layers of chocolate Moenkopi and pastel mounds of violet, lavender, and gray ash follow beneath. The Navajo name for this phenomenon translates as "sleeping rainbow." In the middle of the park, eleven miles by dirt road from the Visitor Center and the small Park Service community, was a guest ranch owned by Lurt and Alice Knee, also named the Sleeping Rainbow.

Although at the time we first visited Capitol Reef we did not even dream it, the Sleeping Rainbow was to be our home for almost four years. A year later, during a second visit to the park, we met Lurt and Alice Knee and ended up staying. We started out caretaking but eventually, we leased the guest ranch from them and opened it up as our own business.

The "ranch" was actually a long motel unit with a spectacular view that perched on the edge of a small mesa at the confluence of three canyons above Pleasant Creek. It included a log lodge, a house trailer, a cabin, and various out buildings. Alice's Arabian horses galloped through green pastures below. Where it temporarily breaks free of its narrow canyon confines above the Sleeping Rainbow, Pleasant Creek makes a small oasis of tall grasses, cottonwoods, and sagebrush. The profusion of petroglyphs, pictographs, pottery shards, arrowheads, and ancient Fremont Indian granaries that are found in the area testify that the creek has been "pleasant" for a very long while. A hay shed, some stalls, and a tack room clustered under the shade from a huge stand of old cottonwoods that once sheltered the Ephraim Hanks polygamist ranch.

Up the creek was a room carved into the rock wall of a canyon with a weathered old door frame built around the entrance. Legend had it that ol' Ephraim and his first wife spent their first winter on the land in that rock room. It is said that the wife became desperate and delusional at one point and Ephraim, fearing she would wander and get lost, kept her in that stone cell by rolling a stout boulder in front of the door when he left each morning.

True or not, the Hanks were long gone by the time Lurt Knee bought the place in 1939. He was originally from Colorado but his dad was killed while working on the Silverton to Durango train and the family wandered to California. Lurt's sister married Harry Goulding who had a trading post in Monument Valley. Harry got tired of trying to make a living off of the Navajos alone and went to Hollywood where he camped out in John Ford's office until the director agreed to see the tall cowboy who wouldn't go away. Harry had some photos of the unique landscape around his place. As a result, Ford and other directors were soon making movies in Monument Valley and using Goulding's Trading Post as a staging area. The lesson was not lost on Lurt who hunted for a place of his own and found it on the border of what was then Capitol Reef Monument. When the monument became a full-fledged national park, Lurt found himself within its boundaries, an ideal situation that gave him the competitive edge over all other local tourist accommodations.

Still, despite his scenic location, he struggled to make ends meet. Utah's redrock deserts were still relatively unknown in the fifties. The hordes of German vacationers, American mountain bikers, and camera buffs you can find on any sunny day today had not yet shown up. So, he tried uranium mining briefly but the mine went nowhere and his first wife left him. Not to be deterred, he eventually found Alice who had old Philadelphia money and an urge to live in the wilderness. She was a tall and big-boned woman with a square face and a high-pitched voice that seemed out of place in such a handsome body. When Lurt met her she was acting as a privately sponsored one-woman Peace Corps to the Navajo nation. When we told Alice that Linda was pregnant she smiled and informed us that although we were seventy miles from the nearest hospital we shouldn't worry. She had delivered two Navajo babies with "nothing but a sharp stick." We cringed, took note, and made sure the gas tank of our car was always full.

Winters were idyllic. In the morning there was time to hunt for cougar tracks in the fresh snow before feeding horses, checking for frozen pipes, and chopping wood. In the afternoon, our chores done, we would hike the trails of Capitol Reef under a crisp cobalt blue sky. By evening, we would begin the ritual of loading the lodge hearth with piñon and juniper logs to burn for warmth and entertainment. There was no television and only a ham radio at the ranch. Sometimes snow or mud would block the lone road from the ranch to the park border and we would be isolated for weeks at a time.

Early spring meant it was time to clear and burn brush away from the irrigation ditches that carried creek water to our garden and pastures. Drifts of tumbleweed that the wind had corralled behind the garden fences over the winter had to be removed so vegetables could be planted. Guest rooms had to be cleaned, repairs made, supplies purchased, advertising lined up, and reservations made. The late spring, summer, and fall were filled with long hours entertaining guests, shopping, cooking, cleaning, and pulling weeds in the one acre garden where we grew enough food to feed ourselves and the daily load of hungry guests. Even when the days were long and dictated by a rigid schedule for cleaning rooms and preparing and then serving meals, we could still squeeze in time to hike up canyon to splash in the creek

Upper Pleasant Creek, 2004.

and cool off. The nights still pulsed with the sudden light of meteors arcing across a crystalline sky.

When our first child was born, we added tending the baby to our list of cares. We could dote on Brian during the winter in our warm cabin, or carry him with us in a tight yellow sling as we climbed the narrow passage up the cliff wall behind the ranch on our daily trip to a stand of old trees we called "the bonsai grove." There we would look out on Mount Ellen in the distance, where a herd of buffalo still roamed free. We would count our blessings and wander home past Anasazi Indian carvings and a nest of peregrine falcons. Hours passed like prayers, quiet and hopeful. We were humbled and awed by the beauty and mystery unfolding around us.

During our years in the redrock wildlands, we learned compelling and fundamental lessons that we had missed during our modern education while immersed in the American way. For the first time in our lives, the connections between our bodies and the water and soil that nourished us appeared short and simple. Back in the city, water was an abstraction. It poured magically from a tap on an underground network of pipes that drained a distant reservoir I had never seen and probably couldn't locate offhand. To keep it flowing, I merely paid a bill. At the ranch, our drinking and irrigation water gathered on the Aquarius Plateau and tumbled down the face of Boulder Mountain, our view to the east, where it fed the riparian oasis under our window. I was responsible for clearing the irrigation headgate and ditches of flood debris and maintaining the pump and pipes that brought water up to a cistern under the lodge. When the pump failed, we literally carried our water each day. One winter we melted snow for cooking and drinking when a pipe between the springhouse and cistern froze.

Our water supply was a daily concern and the cycles that replenished it were at hand and sometimes dramatic. In the summer we could watch thunderheads build and burst, then wait for a flash flood to rumble down Pleasant Creek an hour later. Days later, a carpet of fresh growth would fill in the flood path and deer, bees, and birds would seek the tender grass and wildflowers that embroidered the flood's green wake. Over time, then, if we were attentive and patient enough, the stone-loaded percussion of a flash flood would be followed by a chorus of birdsong and bee buzz.

In eastern America where water is everywhere, water's life-sustaining role is hidden in the lush background. In the baked desert lands of the West, water's gift becomes foreground that is underlined boldly. As any desert dweller will readily attest, water is life. Humans, too, are fluid creatures. I do not mean that we are graceful and agile, though some are, but that we blister, bleed, urinate, salivate, sweat, and cry. Our bodies are a community of fluids. While living on the ranch, the ways water reached our habitat from the sky and how we incorporated it became obvious.

Likewise, years of growing the corn, tomatoes, and melons that attracted and pleased hungry guests taught us that the nutrients our bodily fluids carry arise from soil that was once leaf, limb, stone, root, bone, carcass, carapace, and flower. It was churned, swallowed, and excreted by worms, ants, mites, millipedes, beetles, and bacteria so it could feed the plants that we tended in our garden that, in turn, used the generous chemical energy of the Sun to make food available to us and our wilderness pilgrims. Under the Sleeping Rainbow, our continual bodily communion with the whole wide world was a biological fact of life and the walls we had built between the personal and the planetary during a previous education melted away.

We made a good life under the Sleeping Rainbow but it was not the life I expected to lead. I had lived in big stimulating cities like New York, Boston, and Cologne. I thought I would teach or become an anthropologist, hobnob with intellectuals and engage in fascinating conversations. Instead, the canyons and desert spoke to me and I stayed to listen carefully. I found out "what the Hell" Bill was thinking. My life, too, was drawn and turned by the land.

The most important lessons of my life were learned there, under the Sleeping Rainbow. Those lessons became the lens for the next phase in my life as we moved to northern Utah and lived on the rim of another desert, the Great Basin, where we raised our kids and made our modest careers. We inadvertently moved from the grandest wilderness area in the continental United States to the most extensive environmental sacrifice zone in the nation. In an odd twist of fate, lessons learned living in the wilderness gave me the insight and resolve I needed when I found myself on the rim of apocalyptic ecocide.

The stark contrast of those experiences revealed to me how our bodies are grounded in ecosystems and how we all live downwind and downstream from one another. Those lessons made me understand how the collective decisions we make about what we allow into our air, water, and soil are translated into flesh and blood and living daily experience. Eventually, I also learned how to defend the land that generates our health and well-being from those who would trade that natural health for wealth and power. My struggles carried me into capitols and corridors, television studios, and meeting rooms. But no matter how far into the distant political arenas my journey took me, I never really left the Sleeping Rainbow. I will hold its gifts close to my heart forever.

# 27

## THE NEW FACE OF NATURE

### Utah Valley University Capitol Reef Field Station Students

*The National Park Service bought the Knees' Sleeping Rainbow Ranch in 1978, but Lurt and Alice were granted life tenancy and continued to live there as the buildings gradually fell into disrepair. After Lurt's death in 1995, Alice (confined to a nursing home) quitclaimed her rights to live at the ranch in 1996. Utah Valley University partnered with the park to create a solar-powered field station at the site, which opened in 2008.*

*For several years, writing instructor Kiri Manookin led a four-day Wilderness Writing Workshop at the field station for Utah Valley University English language learners. Most of these international students had never encountered such country, what one called "the new face of nature," and the field class challenged them to apply their developing "English writing ability in a place that stimulates creativity." Manookin saw her international students react to Capitol Reef with "awe, stunned by the landscape. Being there is like nothing they have imagined."*

*She asked each student "to respond to a prompt about the intrinsic and inescapable connection between people and the land." Here are three of those responses—with the raw energy of field notes, before extensive rewriting could smooth out their English grammar and make them less immediate. "Read on," says Manookin, "and you will see, as I do, how new connections formed on this trip between human and earth and the heavens inspire this group of students, and you will sense in them a greater compassion for life."*

## A Family of Rocks (2014), by Salma Hakeem (Saudi Arabia)

*Salma Hakeem came to Utah Valley University from Saudi Arabia in 2013 to study English as a second language. She then attended Arizona State University, to enter the Women and Gender Studies program.*

First time I arrived to Capitol Reef, I was so surprised with the mountains, and when my teacher started to teach me about the rocks' names and about the reasons that made them in incredible ways, I started to be more interested and caring about them. Yesterday we had a little hike. I could touch Wingate rocks. When I saw them first time they looked so smooth and soft, but when I touched them they were the opposite. I felt like they have special personalities. They are soft, but they are still strong.

Today we went into harder hike. In the beginning, I felt tired but I could not stop looking at the mountains. Each time I look at them I feel more about them, I feel appreciated, I feel like they are protecting the ground. They don't let the ground move. So as much I look at them I feel more about them. I feel like they are a free army, and they are best friends to each other. Even when they have different colors and looks, they unite to love their home and protect it, which is the ground.

I felt like they were welcoming us; they love their visitors as long as they don't skip their limits. They are tough and dry, but they were nice with us, they didn't let any one of us fall off them because we were polite.

When we finished the hike and after say goodbye to those beautiful military mountains, we ride the car. While most of my friends were sleeping and checking their photos, I didn't have enough of watching these beautiful mountains. I saw something, like they were smaller rocks in the ground, I felt like they are their kids with different colors and types as their parents. They were playing with each others, and the huge mountains were watching them very peacefully and safely. I loved those soft and beautiful and strong family.

## Moenkopi Grandpa (2017), by Tetiana Novikova (Ukraine)

*Tetiana Novikova is a "passionate software engineer, loving daughter, and sister," and an international student from Ukraine with a BS in system analysis. At Utah Valley University, she studied animation and game development.*

Sunset, Panorama Point, 2006.

I feel young. Sitting here in the Capitol Reef National Park, I feel very young! I realize I am not only looking at the stones, cliffs, or rocks. In front of me, there is the whole history of millions of Earth years. It is impressive that in one place I can see different layers of land created a long, long time ago. Millions of years! It looks so steady and powerful, but at the same time the waterpockets make it look weak and fragile. The place is covered in silence, and only the strong wind sometimes threatens to ruin the quiet.

Standing on the 200-million-year-old rock inspires me and makes me wonder about the future because I am one of the creatures who lives on this planet now. I am the next layer. I am the new generation that can make a difference on the planet, and all the previous layers are like grandpas for the younger ones. They carry history, wisdom, experience, failures, and successes. They help us understand the world before our appearance on the planet. They teach us how to appreciate simple things and fill our lives with peace and love.

Long after this generation is gone, what is the Earth going to look like in 200 million years? Here I can see that time flows really slowly: a thin layer of cliff is a long period of time. After seeing the variety of layers and the immense contrast between them, sand lithification, and thousand-year-old cryptobiotic microorganisms, nothing seems impossible! I feel young and am inspired and ready to be part of the great future :)

### A Land of Rest (2017), by Kadidia Sabou Doumbia (Mali)

*Kadidia Doumbia was born in Louisiana but raised in Bamako, Mali, where as a teenager she started writing short poems. Desiring to improve her life, she came to Utah Valley University to study information systems.*

Here is a land of rest.

Here, I can hear those birds and understand that they are singing of happiness and pride.

In a moment, I do not hear anything. This "anything" does not mean "nothing," but it is a mute song from trees, cryptobiotic soil, sand, plants, lichen on rocks and the Waterpocket Fold. The land is calling me to forget my bad emotions, the bad feelings that I am hiding inside somewhere in my body. The sound of water glowingly flowing is reaching all my sensations.

Actually, it is sunny, and it is incredible because my body no longer hurts being on that warm rock. Generally, I do not like the sun because it hurts; on the other hand, my body here is more exhilarated and comfortable and wants me to stay longer here.

Nature is a place where I can relax my body. Nature is what my body is seeking for. Nature is everything that surrounds humans. Nature is like a mother giving life to a baby. A human can't live without nature because it is the source of how we live. We live by breathing the air, drinking water, and eating foods provided by the environment. All this innocent nature exists, and staying around it is making my body magnetized to the ground.

However, society is a cause of every problem like cheating, hurting, killing, and many other bad things. The difference between here and the city is "awareness," "true love," "enormous." Here, I can no longer be unsatisfied by who I am because my soul belongs to the ground. In society, my head is full of questions; they can be stupid, deep, and sad. Nonetheless, my mind is empty here: no sad feelings, no deep thinking, and no sad tears flowing from my eyes.

Everything here is natural, nobody can fool himself here.

I am getting in love with this environment—the shape of the desert varnish on rocks, iron oxide rust, Boulder Mountain, the form of the Waterpocket Fold attracting me. The formation of those lithifications show what the environment is capable of and how powerful it is.

I envy these trees and stumps; Native Americans were very lucky living here.

In the opposite world when you are under the sun, you want the cold shadow, and when you have the cold shadow, you want the sun. Everyone wants to go higher in getting more money. For example, if someone has a big house, he or she wants the earth in the future. In this natural world it is different. I do not want more but less. The trail here intrigues me and seduces me to go deeper like a baby in front of candy. The volcanic rocks look like the regular rocks, but through my connections with nature I realize they are not beautiful, but special, atypical, and incredible.

Everyone who can understand nature knows the real true life.

Here is a natural world.

# 28

## THE O-BAR DRIVE (1990)

RAY CONRAD

*Ray Conrad has lived in Wayne County for nearly forty years. Artist, musician, chef, he began writing cowboy poetry, publishing mostly in the local newspaper. Over the years he has become the unofficial poet laureate of Capitol Reef country. This poem about a cattle drive through western Wayne County, along the base of Boulder Mountain, down Pleasant Creek through the Waterpocket Fold, south along Notom Road to the Burr Trail, and on past the Henrys into the desert comes from his collection* Fencelines: Poems about Cowboys & Country Folk, Kid Groups & Congress, and Other Rhymes for Our Place & Times.

Up along Storm Mountain the quakes are going brown,
And on the slopes of Hilgard the leaves are drifting down.
The cow-herds in Sheep Valley, and on the UM grass,
Are moving toward the fences, 'cause winter's coming fast.
You hear the riders whistle as they push 'em down the hill,
Cussin' them that straggle, as some old sisters will.
They crowd 'em past Mill Meadow and into Fremont town
Where ranchers cut their own stuff out and shove the rest on down.

The calves get cut and branded, earmarked and shot and all,
And when they're weaned there's two nights sleep that's lost, the way they
bawl.
Those calves can't seem to figure it. Their new world's gone so strange.
But their mammies know what's happening. It's time for winter range.

The old chuckwagon gets all stocked and swept out real nice.
The cook, he kills a good fat ewe and packs it down in ice.
The horses get new shoes on, 'cause it's November Third,
And we're headin' for the desert behint the O-Bar herd.

We take out from the home ranch quite early that first day.
Down around the airport. Them old cows know the way.
Along the Bicknell Bottoms, then swing 'em to the right,
Through the town of Teasdale, to Fish Creek for the night.
The tin chuckwagon's waiting. The stove is good and hot.
There's biscuits in the oven, and sheep ribs in the pot.
Sourdough and mutton! That's the menu that's preferred,
When ridin' to the winter range, behint the O-Bar herd.

We maybe sip a toddy before we set and eat.
With ice and mountain water that whiskey sure tastes sweet.
And after we get fed good, and set and gab a spell.
We hit the blankets early and prob'ly sleep quite well.
First light in the morning, the cow-boss gives a shout.
We feed the ponies, then ourselves, and turn the cattle out.
They slowly graze through Grover town. We follow without a word.
We're on the way to Hansen Creek, behint the O-Bar herd.

When the sun gets off the ridgetops we got 'em going good.
We ride off Miner's Mountain through sandy piñon woods.
We lunch in Happy Valley. A sandwich and a beer,
Then down the gorge of Pleasant Creek. It's rough and rocky here.
On to the Sleeping Rainbow, a brushy, dusty ride,
But that chuckwagon's waiting there, with food and drink inside.
We move the cow-herd through the fence, then jaw 'til dinner's served.
Camped in a red-rock canyon, excortin' the O-Bar herd.

Another gourmet supper, with sourdough, of course,
Then wake up soon to breakfast, for rider and for horse.
Down the creek and up the slope. The view from here's a thrill.

Dwight Williams, Wayne County rancher, demonstrating an electric branding tool that allows for freehand branding, Capitol Reef Harvest Homecoming Days, 2015.

Watchin' them old heifers stringin' down that big long hill.
Then south along the Notom Road. We're going a long, hot way,
But Cooky's waiting by the road to feed us lunch that day.
A big old mutton samwich will make your belly purr,
Whilst heading down towards Bullfrog, behint the O-Bar herd.

We camp near Sandy Ranch that night, along Dogwater Hill.
No doubt we have a toddy, as thirsty cowhands will.
Some 'locals' come to visit camp, and Cooky feeds us all,
And then we gab around the fire 'til coyotes start to call.
Tomorrow is a long, long day. We move 'em out at dawn,
So sleep quick, 'cause the night is short, and we must soon be gone.
If one word could describe this day, 'dusty' would be that word.
Straggling over Bitter Creek, behint the O-Bar herd.

We eat our lunch on horseback. We ain't got time to stop.
The cows are getting thirsty. The ponds don't hold a drop.
There's other cattle grazing here. We dasn't let 'em mix,
'Cause having to stop and sort 'em out would put us in a fix.
We get to the fence what they call The Post.
That's where we're gonna stay,
And when we strip them saddles off we know we've had a day.
We're all about to choke to death. Our eyes are red and blurred,
From eatin' all that desert dust behint the O-Bar herd.

The horses get some barley. They just want to hang their heads.
Us riders have a drink or three, and then we all get fed.
We maybe play a poker game, and swap some windy tales,
Camped in the rocky desert, near to the old Burr Trail.
Then onward in the morning, through cactus, rock and sage,
Movin' cows the way they done in an older, slower age.
Coming close to Egganog Spring. There's water there for sure,
Getting near to the winter range, behint the O-Bar herd.

We got to Egganog early. It started in to rain.
It's all blue mud in that big hole, and that can be a pain.
We pushed the cow-herd clear on through to the Clay Point gate that day,
'Cause if Saleratus Creek got full there'd sure be hell to pay.
When we got back that crazy creek was runnin' deep and wide.
The whiskey and the food and beds was on the other side.
I kicked my pony through the flood, employin' some fine cusswords,
And the cows were safe on the winter range.
The whole damn O-Bar herd.

Every cow had made it there. Not one of them had died,
But we was stuck at Egganog Spring until the mud had dried.
Which took about a day or two, and we loaded for Fremont town.
We'd finished our assignment of chasin' cows around.
It seems we'd drank up all the booze and ate up all the food.

It's way past time we had a bath. We all smelled kind of crude.
That trip was long and dusty, but you can rest assured,
I'd like to make that ride again, behint the O-Bar herd.

But there ain't no O-Bar drive no more. It's a footnote to history.
No one has the time, no more, to ride that far, you see.
They have to tend their day job and try to make a buck,
And the cows hitchhike to Hansen Creek on Dewey Woolsey's truck.
The old chuckwagon gathers dust. It's got no place to go.
No more Ten High cocktails. No mutton and sourdough.
But often I think on an autumn day, say about November third,
How fine it'd be ridin' south again, behint the O-Bar herd.

Park headquarters and old Fruita, from Rim Overlook, 1992.

# PART VI

## FRUITA

### Verdant Village, Sweet Sanctuary

Capitol Reef National Park runs for seventy miles along the Waterpocket Fold and protects nearly 241,000 acres of the Colorado Plateau. But for the majority of visitors, Capitol Reef means "Fruita." Tucked between the vermilion Wingate cliffs where Sulphur Creek and the Fremont River join, the historic village's bountiful orchards and emerald fields enchant summer visitors worn down by austere badlands.

William Byron Pace reached the site of Fruita in the summer of 1866, when he led a company of Mormon militia through Rabbit Valley (which he named) and down the Fremont, searching in vain for the Utes and Paiutes who had triggered the Black Hawk War. Pace and his men commented on the "impenetrable Rocks," found no trail through the cliffs, and withdrew. But their foray mapped new territory for Mormon settlement.

Franklin Young, Fruita's first Anglo settler, admits that the place was too lonely for him; he didn't last long. Former chief park naturalist George Davidson gives us the background history of the settlers who stayed; he highlights two residents, Neldon Adams and Dewey Gifford. Clay Robinson remembers with affection his childhood years at the Fruita school.

Many reminiscences of Fruita end on a bittersweet, elegiac note. Writer Jen Jackson Quintano captures this longing in the jars of Capitol Reef apricots she keeps under her bed.

Wallace Stegner also loved Fruita. In 1977, the distinguished writer of the American West visited southern Utah to gather material for his essays in *American Places*. Working with these fresh impressions, Stegner captured the feel of that "lost village" of Fruita, a "sanctuary" for "enthusiasts with the atavistic compulsion to hole up in Paradise." Renny Russell gives us another view of those one-of-a-kind originals hiding out in Fruita, a remote place in the middle of the twentieth century.

Capitol Reef's first superintendent, Charles Kelly, a complicated man, deserves extra attention. Historians Gary Topping and Jonathan Thow reveal Kelly's contradictions.

But first, a song from Kate MacLeod, as invocation.

29

## PICK, PICK APPLES (2016)

KATE MACLEOD

*Utah composer, singer, and fiddle player Kate MacLeod was the Artist-in-Residence at Torrey's Entrada Institute in 2016. The people who pioneered and lived in Fruita and its orchards inspired her to write this song.*

Along the Fremont river
Tucked into a canyon
Grow the orchards
That I ran through as a kid

That was everything to me
When I was simple, wild and free
Before I knew the world was so big

The roads were rough and people rarely
Came our way, but when they did…

CHORUS
Oh, it was pick, pick apples and story time
We knew how to pass the time
In the middle of nowhere
Where we lived

I worked in the sunshine
Like the people who left behind
Their baskets and petroglyphs

Fruita was our Eden
Where we grew thousands of trees
Between the river and the cliffs

And travelers came from everywhere
In the big wide world
Just to see

CHORUS
Oh, it was pick, pick apples and story time
We knew how to pass the time
In the middle of nowhere
Where we lived

And I am who I am
Because of my home in that canyon
And my kin

CHORUS
Oh, it was pick, pick apples and story time
We knew how to pass the time
In the middle of nowhere
Where we lived

# 30

## THE JUNCTION (1909)

### FRANKLIN WHEELER YOUNG (1839–1911)

*Mormon patriarch Franklin Wheeler Young helped settle the LDS frontier, from Bear Lake to the Grand Canyon. With his two wives and sixteen children, he applied his pioneering drive to Wayne County for two decades. On an 1883 exploring trip, he camped at the junction of Sulphur Creek and the Fremont River. Intrigued, he returned to build a home in 1885.*

Early in June [*1884*], I got home and looked after the farm the best I could through the summer, but the early frosts took my crop again, and I became discouraged. During the next winter, I went down the river to the junction of Sand Creek [*now called Sulphur Creek*] with the Fremont River, made a location and dug a water ditch. In the spring of 1885, I planted out an apple orchard, with a few peach and other trees. I built a house there, removed Nancy and her effects down there, and sold out my land at Rabbit Valley.

Later on, I bought into Teasdale and located both of my families there, selling out at the Junction [*now Fruita*], because I was alone there and at that time, no one could see anything there worth having.

# 31

## ECHOES OF CHILDHOOD IN OLD FRUITA (2004)

CLAY M. ROBINSON (1920–2013)

*Clay Robinson worked professionally as a writer and editor for in-house agricultural and fishing publications. He brought these writing skills to bear when he cowrote a collection of his childhood memories of Fruita with his brother, Max—* Echoes from the Cliffs of Capitol Reef National Park*—from which this story comes.*

From the towering red and yellow cliffs of Capitol Reef National Park come echoes and pleasant memories of my childhood of nearly eight decades ago. I am melancholy because those happy days are gone forever.

Most vivid of all in those memories, are Mama and the little one-room, sawed-log Fruita school where she taught all eight grades in the years 1925–27. In her first year her pupils ranged in ages fourteen years down to me at four. She had no other place to put me, so I became a first grader, and the school mascot.

My brother, Max, grown up at six-and-a-half that fall of 1925, also became a pupil of the Fruita school. He was my pal. Had I been left at home alone in the old rough-board cabin where we lived only a short trot from the school I would have felt forsaken. So I was glad when Mama enrolled me at her school.

We small boys, along with our mother—and other people in the small tranquil community—were happy. To us the quaint village of Fruita would always be there. Its life would always be unhurried. The early sun would rise to light the red, yellow, and white cliffs to jewel-like splendor. And on moon-

Behind the Fruita school; Dewey's and Nell's sons, Lloyd and Fay Gifford, sit in the center, 1935. Courtesy Capitol Reef Natural History Association.

less nights a myriad of stars above the unique Mormon enclave would shine so brightly you could look far into the many folds of heaven. On occasion, under the glow of a full moon, faint shadows of the cliffs would spread over the village. We could not envision a time in the future when Fruita would become a ghost town in the heart of an intriguing national park.

In those old days, my mama, Hattie Mulford Robison, was the smartest and prettiest teacher in the whole world. Well, maybe not any smarter than my daddy. He, too, was a teacher, and principal of the big three-room school at Torrey, twelve rugged and winding dirt-road miles to the west.

As school mascot I enjoyed many privileges. Older kids envied me, for I randomly chose my lessons. Nature classes offered the most fun. In front of the school I'd tease the ants until they'd sting me. Then along the path to Sulphur Creek I'd play with grasshoppers and toads. Occasionally I wandered into Dicey Chestnut's bountiful orchard. There I'd pick a tree-ripened peach, apple, or pear. I'd eat until juice flowed down my chin and attracted bees and pesky flies on the wing for sweets.

Aunt Dicey (she was not actually my aunt, but everyone called her aunt) didn't mind my eating the fruit from her trees. She, a generous woman, loved children. She raised five orphaned children besides the four of her own.

In that first year of school we kids held as our hero, Jack Dempsey, a young Mormon who drew globe-wide attention as world's heavyweight champion boxer. We boys wildly swung fists in preparation to become champion pugilists, just like our hero. We'd take a poke at anything and anyone just for practice.

Mama, at only twenty-nine that basic year, experienced her first teaching in a school. She had been going summers to the far-off university to get her certification to teach. It was then she left Max and me for long months with Grandma Mulford on the old Vermillion ranch.

We missed Mama then, and tried to figure out ways in which we could make the 200-mile trip to see her.

We constructed an airplane out of an old hay rake. Surely that contraption would fly us to see Mama.

When the hay rake failed to take off, Max got Uncle Con's new hard-twist lariat from the shed. He roped a pig. We would ride that pig to see Mama.

But the pig would not cooperate. We held tight to the rope and almost got dragged off the fence into the pig muck before letting loose.

Then came the awful confessing to Uncle Con. Our faces burned with shame as we watched him retrieve his soiled rope from the pig's neck.

Today, when I go back to Fruita (that's what Capitol Reef National Park will always be to me), I perch myself high on a cliff. I look across the way that once was the verdant village of Fruita. In view are the old Pendleton barn and the Dewey Gifford house. But, most vivid of all, is the little, rustic, one-room school. The school is now a shrine.

In my mind I still see Uncle Cass' fruit orchard to the south and west along the river. The orchard is gone and, unfortunately, so are his old cottage and the unusual rock-lined cellar, built into a large red-clay knoll. There Aunt Marie stored produce, meats, and dairy products. It was cool in summer and prevented freezing in winter.

Also, in my mind I see the Chestnut home nestled against the big purple and black mesa. Like other reminders of yesteryear, it, too, is gone.

Apricot harvest, Fruita orchards, 1994.

Replacing these remnants of the past are camp grounds and parks. I overlook them, for I still see Fruita in the days of my childhood. My throat throbs a bit, for I wish these old reminders had been saved by the Park Service to give tourists a better picture of a society now extinct.

From my lofty vantage point, I listen intently. From the echoes of the cliffs I hear those familiar voices of the past: Uncle Cass, from his ranch house up the valley, calling to his neighbor, Dewey Gifford, down close to the confluence of the river and the creek. They had no need for telephones.

Their voices readily pierced the three-quarters of a mile and echoed like thunder in the enamel-smooth red cliffs.

And I can still see and hear Mama stepping out onto the sandstone-slab deck of the school. She is ringing the hand bell and calling in her little flock of boys and girls.

All those echoes still resound today, locked within the concave chambers of the towering, vertical cliffs.

...There had been many teachers before Mama took over. And many teachers after she quit teaching. The little Fruita school continued to turn out good, knowledgeable students up to its closing in 1941.

Among the students of that little school one finds today a retired general of the army, a range management specialist with a master's degree (my brother Max), and various successful business people and professionals.

I remember the last time I attended a social function in that tiny school house. Yes, it was used on Sundays for church meetings and Sunday school, and for special community socials. That last time was in 1940 at a farewell party for my cousin, leaving to go on a mission for The Church of Jesus Christ of Latter-day Saints, the Mormons.

Familiar faces filled the room. Kids who had gone to school with me had grown into adulthood. And their parents had grown old. But all of us, young and elderly, played the same old games of yesteryear. We sang the same old familiar songs that we enjoyed just as we had back in the days when I was a first grader.

Today, as I gaze down on the tranquility of the little dell that once was Fruita, Utah, I see the green foliage-lined creek and the river. I can still hear the echoes of the tremendous, thundering, boulder-rolling and pounding floods that overflowed the creek channel after a cloud burst in the Upper Country. Those floods always scared the heck out of me.

I closely scrutinize the area just below the black boulder-strewn mesa near the river. There, in my reverie, I see Old-man Jorgensen using a horse hitched to the end of a long pole extended from a crude homemade cane press.

The horse trots around and around on his well-beaten path to power the press. Juice spews from the grinder as Mr. Jorgensen feeds in the sorghum stalks. He made good sorghum, if you like sorghum [*a grain that yields a sweet molasses-like syrup*].

Most of the people I knew in those days have faded away, gone to explore and settle other frontiers in The Land of Hereafter. My mother and father; Grandma and Grandpa Mulford; Uncle Cass and Aunt Marie Mulford; Dan Adams, who rode the big sorrel gelding; Old Mr. Jorgensen; Dewey and Nell Gifford; Dicey and Will Chestnut, and their children, Clarence, Lucille, and Jay; Valentine Oyler; Marin and Cora Oyler Smith, and Al Chestnut, along with many of his sons and daughters. And there were many others, from my time, and before. They are all gone from the pleasant little dell that once was Fruita. But their voices, their laughter and their weeping—from times both good and bad; happy and sad, still linger in the echoes from the red and yellow cliffs of the Capitol Reef.

# 32

## POVERTY FLATS & PARADISE (1986)

GEORGE DAVIDSON

*George Davidson served as chief of interpretation on the park staff at Capitol Reef in the 1980s and 1990s. He loved local history and collected many stories from Wayne County old-timers. This excerpt comes from his history of Fruita,* Red Rock Eden.

What was it like to live in isolated Fruita before it was discovered by the National Park Service? To find some of the answers, we need to visit an earlier twentieth century Fruita through the eyes of one-time residents.

Not surprisingly, views differ. One relative latecomer, G. Dewey Gifford, had very pleasant memories of his more than forty-year residency and referred to Fruita as a "paradise." On the other hand, G. Neldon Adams, who came to Fruita as a boy in 1914 and whose family left just before Gifford's arrival in 1927, had vivid memories, too, but was glad to leave. He remembered Fruita as "poverty flats."

No doubt much of the disparity in these recollections can be explained by the fact that Neldon Adams was a teenager here, kind of an involuntary, hard-working prisoner in a social isolation that he yearned to transcend. Dewey Gifford, on the other hand, came here as a newly-married, resolute, and self-sufficient man who was seeking to make a home and prosper in a landscape he found beautiful and challenging. The differences in their perceptions, perhaps explicable, are still interesting.

Fruita in its earliest days, around 1900. Inglesby photograph collection, Special Collections, J. Willard Marriott Library, University of Utah.

## Poverty Flats

Neldon Adams' father bought old Joe Cook's twenty-odd acres in 1914.... The Adams family was large—not unusual for Utah Mormons—and consisted of mother Harriet; brothers Myrcus, Farrell, Orville, Clifford, and Dee; and sisters Reeda, Novella, Fauntella, and Hazel.

Neldon's father, Andrew, worked away often as a sheepherder to earn the cash that was otherwise nonexistent in a part of America where bartering was status quo (and is still widespread). Much tedious farm work fell to young Neldon on the twenty-acre farm that was about half orchards and half pasture and hayfields. One job he particularly despised was the frequent and backbreaking cleaning of irrigation ditches that rapidly silted in, especially the ditches from aptly-named Sand (Sulphur) Creek. It was the massive silting-in of such ditches by floods in Caineville that broke the spirits—as well as the backs—of some of the toughest human beings on the American frontier.

Neldon remembers that he was doing "a man's work" when he wasn't big enough "to reach the trips on the hayrake" (a horse-drawn machine that gathered cut hay in rows). Besides school, he milked a few cows every day, cut wood, hauled water barrels up from the creek and—in season—picked fruit. He helped his father haul the fruit up the valley in their Studebaker wagon to sell or barter for grain and cheese.

Neldon was often bored and got into more than a little mischief. One spring, all the cats in Fruita seemed to come down with a mysterious affliction that resulted in their tails falling off—"bobtailing"—or so it was first thought around town. It wasn't until Neldon's coconspirator—Jessie Richardson—broke a vow of silence that the people in Fruita found out that he and Jessie had been chopping off the poor felines' tails with a hatchet. Neldon fondly remembers his mother's declaration that "if it wasn't so cussfired funny, I'd skin you alive."...He once climbed to the roof of the schoolhouse when the stove was fired up and threw black pepper down the chimney, causing convulsive sneezes.

As he matured, Neldon's thoughts turned to girls; there were a few nearly his age around Fruita. He especially remembered "Tine" Oyler's attractive daughters Cora and Carrie. The sudden, tragic death of beautiful Carrie, hospitalized for appendicitis, stunned the community and was a hard blow for her family.

Teenagers often swam in a Fremont River hole in back of the still-standing Gifford house. They often held dances and enjoyed candy pulls and chicken roasts. The boys were responsible for catching and killing the chickens, the girls for roasting them. Neldon reports the adventure of a foray up the valley to obtain chickens; once he and two pals liberated some from captivity, and they really had a feast. There wasn't much in the way of music for these parties—just a harmonica or a poor fiddle—although the Oylers had a player piano and the Adams family owned an Edison cylinder phonograph. Even into the mid-1920s, no one had yet acquired a battery radio.

Going up in the valley was something you could look forward to, remembered Neldon, "you could even buy a candy bar" from Bishop Pectol's general store in Torrey. Neldon, who was always small, liked to attend the big dances there, but had to make his mark with his fists until the local boys no longer tested him.

Neldon knew well and worked for all the farmer-moonshiners around Fruita. The hardest work in making moonshine was the hauling of large quantities of water to dry, cliffside hideaways, although one site—Whiskey Springs—was well supplied. Liquor distillers made some potent shine from corn mash, fruit, and sometimes exotica like "the green skimmings of molasses" when nothing else was available. They also put up enormous quantities of grape wine. All this brought in cash revenues from thirsty sheepherders, cattlemen and others. Legend has it that hard-drinking, traveling sheepshearers consumed by far the greatest quantities of these products. Much moonshine was consumed right there in Fruita, too, according to Neldon.

The valley of the Fremont was so remote that the great care taken to conceal stills and liquor caches was mostly unnecessary. It appears that moonshiners had little to fear from local law enforcement officers. Once, however, their care in concealment paid off. Automobiles were a rare occurrence in Fruita and when one arrived at 2:30 a.m. one morning in 1918 everyone was alerted. It was a huge touring car full of determined state revenue agents. The next day, the "revenuers" found a big cache of the most widely known of the Fruita moonshiners (hidden under the hinged deck of a front porch) but they failed to ferret out the producing stills. It caused quite a stir, but the local moonshiners' operations were rarely bothered again.

Life in and around Fruita was sometimes physically violent, according to Neldon, who often proved he could handle himself. Neldon remembered young Merin Smith, Tine Oyler's hired man, as a nice fellow but no one to be trifled with. Also, heavy-drinking, camping cattlemen liked to fight, especially when "well-oiled," as Neldon put it. Nearby town dances—mostly sponsored by the Mormon Church—very often included a big brawl as a sideshow, unscheduled of course. But the rough stuff was not the deadly kind, at least most of the time.

Neldon remembered one time when a local farmer was insulted by a cowboy during a camp house drinking spree and took off after his rifle. Fortunately for all, the farmer passed out in the creek on the way back from his home. One time Neldon personally witnessed a gunfight over the whereabouts of a woman that left one man wounded. The wounded man was able to drive away in his Model T Ford, however.

People passing through Fruita were always of interest. Besides sheepherders, cattlemen, and sheepshearers, Neldon reported a few strange travelers, especially gypsies. Neldon recalled an incident when an old gypsy woman offered to read his fortune if he would gather a pile of firewood. He did so, but she refused to pay off. When he threatened to tell his father and complained bitterly to the gypsy leader, she told his fortune in detail—really "laid it on thick," according to Neldon.

When he was quite young, Neldon's forthrightness almost earned him a swift kick from a traveler. Two men in a big Pierce Arrow touring car came through wearing dusters, goggles, and gloves—early sightseers. Pointing up the road to Capitol Gorge they asked about the mileage in the peculiar Utah idiom of the day, "How do you call it to Hanksville?" Neldon, taking them literally, responded with the only name he had ever heard for the road and declared loudly "a rough sonofabitch!" Neldon recalled that they chased him into the house without catching him, but they offered to apply one of their black leather boots to his posterior.

## Paradise

Dewey Gifford was a tall, powerfully-built, quiet man who was, for much of his life, a devout practicing Mormon. He and his wife, Nell, bought out her dad (Jorgan Jorgenson) and went on to raise a family of four children—Lloyd, Fay, Dale, and Twila—in a two-story, stucco house near the Fremont River. If any man made a reasonable success of living and farming in Fruita it was "Dew" Gifford. In 1927 his neighbors were William Chesnut, a widower with four children; Al Chesnut; Merin and Cora (Oyler) Smith; Guy Smith; Clarence "Cass" Mulford; Dan Adams; and Tine Oyler.

Dewey had worked in the Wyoming oil fields to earn a stake but he was—soul and sinew—a farmer. He was known as a tireless and thrifty worker, but he had to learn the fruit business mostly by listening to some good advice and by trial and error. On occasion, experience was a pitiless teacher and Dewey remembered losing a whole grove of apricot trees, because he gave them too much water from his irrigation ditches. To bring in cash, he "worked state road" part time for several years.

As far as farming was concerned, the fruit trees were everything. Except for vegetable gardens, the only other crop was hay for the animals, although

Dewey Gifford rescuing a tourist mired in Fruita mud, 1945. Courtesy Capitol Reef Natural History Association.

some sorghum had been grown earlier in the century. Horse drawn equipment was primitive, even for Depression America, with walking plows, heavy wagons, dump rakes, and mowing machines mostly in evidence. Dewey borrowed a sulky plow from Torrey from time to time. All these implements were pulled not by classic draft horse breeds, but by multi-purpose animals used for both saddle and harness—"plugs," Neldon called them. In Dewey's younger days, Fruita residents were acquiring cars and trucks; young Merin Smith had the hottest machine in Fruita, an early V-8 Ford.

...By Dewey's day, Mormons were journeying to Torrey for worship and other activities. In 1900, it had taken two hours by buggy over the steep, twisting dirt road; by 1930, a car or pickup could climb the grade in a half hour. It seems that only a few adult males in Fruita were practicing elders in the Church in 1930 and this appears to have been a significant decline from the turn-of-the-century period. In 1930, Fruita stood outside the mainstream of Wayne County in more ways than geography.

Spring and summer were busy times—first pruning and hauling off cuttings, then irrigating and fruit picking. Women "bottled" foods at a furious pace in the summer to get ready for winter. Fall was a time to slaughter hogs, sometimes a steer, and to salt down meat for the winter. Dew Gifford was one of a few Fruita residents to smoke pork as well as salt it; he inherited a good smokehouse from his father-in-law and used it often. He favored not the abundant applewood of Fruita for smoking, but green cottonwood instead.

In the dead of winter, Nell Gifford sewed most of the family's clothes, except for the overalls worn by men and boys. She quilted, mostly with her two half-sisters in Fruita, Dicey Chesnut and Marie Mulford. Besides mending fences and tending to the equipment, winter was a time of boredom and bouts of cabin fever. People passed around well-read books, played cards, listened to the battery radio and usually went to bed early.

Family outings were much anticipated. It was easy to catch huge trout not too far up the Fremont canyon, and high country deer hunting expeditions were a major source of pleasure (by 1930, not a deer was to be found near Fruita; they had long been hunted out). Occasionally, trips were made to Richfield and Salina, although the road was unpaved almost the entire distance—about eighty miles. In Fruita, holidays were marked by picnics and baseball games on homemade diamonds.

Dewey became and remained an ardent Franklin D. Roosevelt democrat, pretty rare in conservative rural Utah. He and republican neighbor Cass Mulford often had some spirited political discussions.

Dewey's sons went off to war in the early 1940s, surviving service with the U.S. Navy. Dewey continued to do well in Fruita because he never ceased working. As visitors started arriving to see Capitol Reef National Monument he saw new opportunity; Dewey tore down the decrepit camp house and replaced it with a small motel, becoming an innkeeper as well as a farmer. Dewey even took a job with the maintenance force of the park, but in his late sixties he took the cash offered by the government for his farm and he and Nell moved to comfortable new quarters in Torrey, very close to the little church they loved. They were the last of the old timers to leave, and they closed the book on a little-known but fascinating chapter in Utah history.

33

## JEWELED JARS OF MEMORY (2011)

JEN JACKSON QUINTANO

*Jen Jackson Quintano made her home in redrock country for a decade, finding stories and strength in the sweep of southern Utah's canyon country—and writing essays like this tender reminiscence of Fruita. Her book* Blow Sand in His Soul: Bates Wilson, the Heart of Canyonlands *explores the life of the "father of Canyonlands National Park," who also helped research and plan Capitol Reef's 1969 expansion.*

This year, twenty-six quart-sized Mason jars hold the season's offering of sun-soaked apricots, preserved for the months when abundance gives way to want. The jeweled jars glitter like treasure under my bed, the only available storage space in my small desert home. This is my food-based fortune.

I picked this fruit at Capitol Reef National Park, home to some of Utah's most stunning landscapes. The park is also home to the Mormon settlement of Fruita, a place emptied of its residents but still resplendent with their colorful, fruit-bound legacy. The Park Service now tends to trees that ensured life and livelihood for generations.

My jars of apricots hold memories within their matrices of syrup and fruit. A dear friend and I went to the park to harvest amidst the monsoon season's fickle moods. When we arrived in the orchard, so did the deluge. In no time, the Fremont River swelled, the towering cliffs erupted into a chorus of torrential waterfalls, and the orchard flooded in a bubbling refrain of red

Apricots, 2017.

muck. As we picked fruit, we waded through shin-deep mud, enjoying the best of childhood in the process: climbing trees, gorging on syrupy sweet fruit, and covering ourselves in sloppy red earth. We returned home wearing a sticky, earthen residue of summer and joy. This is all packed into my jars. To finally eat the fruit will be to relive the memory. I am mindful that I am storing stories under my bed, my dreams permeated by their sweetness.

These jars are also full of my singular passion for this place. Just thinking about Capitol Reef—and the Henrys and Boulder Mountain—makes me homesick. It makes my heart squeeze. It makes me want to run away right now and go *home*. My heart is tenderly yet tightly held in the jagged teeth of a rock ridge slicing southward through the desert. My heart is sunk deep inside Navajo sandstone, like a Moqui marble. Maybe the forces of time and erosion will loose it someday, or maybe this body will be long gone by the time my heart is freed.

When I am away from the sinewy, sediment-laden Fremont River and its surrounding topography, I feel the pulse of its fickle current in my veins. I hunger for it. Jeweled apricots carry my senses there when my body is far away. Preserving the fruit, I bottle up bits of my ardor, to reclaim in moments of pang and privation. In those jars is the work of hands that have delighted in silken mud, that have caressed sandstone curves, that have cradled collections of stones and feathers and cedar beads. When I eat the work of these hands, I am ingesting moonrise over cottonwood trees, echoing thunderclaps preceding rain, sand and sage-scent carried on the wind. Everything that nurtures the fruit nurtures me. My jeweled jars tell a love story. There is magic in this place.

The apricots carry another story as well: the history of Fruita. The settlement was once full of families whose connection to the land eclipsed that of any mere visitor, no matter how ardent. Their very lives depended on that red earth. The canyon's residents anchored their world to the junction of Sulphur Creek and the Fremont River with orchards, the thousands of trees helping to make meaning of a life rife with floods and scarcity. So accustomed to barter and simplicity was this small community that the Great Depression's lack of cash flow had no effect on it; Fruita's isolation rendered orchards its treasury and fruit its currency. As the country struggled, this settlement soldiered forward as it always had—with a pocket full of faith and a pantry full of fruit.

And then came the designation of the landscape as a national monument. With it arrived tourists, paved roads and the outside entering in—the death knell for a town clinging to a past that modernity had made obsolete. In preserving a landscape, the government had inadvertently evicted those populating the terrain with layers of story. The Park Service bought out the orchards and forcibly evicted those living in the path of the new highway. Many residents recognized there was no other option but to leave; their remoteness had rendered them an anachronism that would crumble beneath the flood of progress. They accepted government money and reluctantly moved on. Most of the buildings were quickly razed to make room for park infrastructure. A raging Fremont could have wrought no greater destruction in this small town. And, thus, a rooted narrative was silenced to make room for a newer one of snapshots, scenery, and short stays. But, as with my apricots, this preserved place is rife with memories. Its preservation has kept the memories alive and made room for new ones. Such as my own.

Today, 2,400 fruit trees live on to tell a quiet tale of communion with place, of space both inhabited and wild. And my apricots connect me—if briefly, tenuously, with just a taste—to this narrative, one beyond the insularity of my own small love story. These jeweled jars hold tales of a time when money was less meaningful than the vagaries of frost and flood, and fruit could build or break one's world.

# 34

## HIGH PLATEAUS (1981)

### WALLACE STEGNER (1909–1993)

*As a seasonal ranger at Capitol Reef in 1975, I had reason to write to Wallace Stegner, asking for permission to use his words on postcards, pairing my photographs with quotations from the likes of Stegner, Abbey, Muir, Dutton. Stegner was right up there in the pantheon of writers about the American West, and his love of Capitol Reef shone in his books* Mormon Country *and* Beyond the Hundredth Meridian: John Wesley Powell and the Second Opening of the West.

*Stegner's responses were warm, immediate, and encouraging. He recalled working to upgrade Capitol Reef to national park status in the 1960s but cautioned me, in jest: "Don't tell any of the local ranchers, or they'd send me letter bombs."*

*He wrote back to ask for leads on local people when he came to Fruita to renew his boyhood memories of the little oasis in the slickrock for* American Places, *"just to see how the present has intruded on their serene isolation." This excerpt comes from what he published after that trip.*

*I now realize that when I wrote to Stegner from Capitol Reef as a young ranger, I had already won him over with my return address. He made a point of journeying back to this favorite place in both life and art. When Stegner wrote his autobiographical novel* Recapitulation, *he set the courtship and consummation scene for his main characters in the orchards of Fruita, by moonlight.*

*Here, "they are in a pocket of green among red cliffs. A dusty track turns off left."*

*"'Here,' the girl says, and the driver swings the wheel."*

*Here, "on that bedroll in the moon-flecked shadows of the cottonwoods beside the guggle of an irrigation ditch under the Capitol Reef."*

Any return is a journey across both space and time. Driving along the top of the low, treeless Awapa Plateau, aiming for the gap between Thousand Lake and the Aquarius, I find my memory as busy as a squirrel digging up acorns. My first trip into this country was in 1924, pretty primitive times, in the company of my scoutmaster and his son. The scoutmaster, an Episcopal missionary to the Mormons, died a disillusioned man, without converts, but in the back country he was an enthusiast. Also he was a pipe smoker. Coming around the corner of Thousand Lake Mountain, as we are coming now, with the Red Gate cliffs on the left and Rabbit Valley around us and the white domes of Capitol Reef showing ahead and the profile of the Aquarius high and dark on the right, he made a sudden, excited gesture with the pipe in his left hand. The wind blew the coal out of the pipe and up his sleeve, setting his armpit afire. His son and I died laughing and were buried just here, where the road dips down toward the canyon of the Fremont and the lost village of Fruita, now absorbed in Capitol Reef National Park.

Fruita used to be one of our favorite places—a sudden, intensely green little valley among the cliffs of the Waterpocket Fold, opulent with cherries, peaches, and apples in season, inhabited by a few families who were about equally good Mormons and good frontiersmen and good farmers. Over the years it has also been sanctuary for a number of enthusiasts with the atavistic compulsion to hole up in Paradise, people who like Zane Grey's Lassiter put a bloody hand to the balancing rock and rolled it down to block away the world.

One was Doc Inglesby, who at the beginning of the 1930s sold out his business, the Salt Lake–Bingham stage line, and came down here to be a rockhound, run a little motel, and listen to the singing of the cliffs. He was a little round gnome with a little round belly and a little round cocker spaniel that he used to carry on the pommel of his saddle when he rode around exploring undiscovered canyons, measuring unnamed natural bridges, and hunting jasper geodes. His passion for the country was as explosive as gasoline: he could scare you to death calling your attention to a sunset or the light on a cliff. I hope the sunsets are good where he is now, and the ledges red.

Another was Charlie Kelly, a considerable Mormon-eater, the author of several basic books on Utah's frontier history, especially its outlaws, and

Doctor Arthur "Doc" Inglesby and petrified wood, Capitol Reef area. Photograph by Charles Kelly. Courtesy Utah State Historical Society.

the first superintendent of the Capitol Reef National Monument. He was a good historian and a good companion. *Requiescat*. He too deserves an Eternity of red ledges.

A third was Dean Brimhall, whose father had been president of Brigham Young University and who had spent much of his mature life in the Department of Commerce in Washington. Related by blood to half the Mormon hierarchy and by marriage to the other half, he was a sadly lapsed Saint, as antagonistic to the Church as he was devoted to the geology and ethnology of the plateau country. In the early 1940s, when he and his wife Lila were visiting us in Vermont, he wished aloud that he knew of some place in Utah as quiet, remote, and peaceful as our Vermont farm. I suggested Fruita, and the next I knew he had gone down there, fallen in love, and bought a piece of land, with a log cabin, a grove of big Fremont poplars, and an orchard.

Before long he built a house, before much longer he was living in it most of the year.

By the time of his death in 1974, Dean had become the greatest student of the plateau region since the Powell Survey quit working there. He knew every crack and canyon, every arch and natural bridge, every petroglyph and pictograph, within a hundred and fifty miles of the Capitol Reef. With great labor he brought in scaffolding and set it up in remote canyons to photograph cliff murals that constitute the finest body of native American art north of Mexico. Some of these Mary [*Stegner's wife*] and I have seen *in situ*; most of them we know from studying color transparencies in Dean's Fruita house—deer, elk, bighorn sheep, square-headed men, hands, the records of how the people who lived in these canyons a thousand years ago responded to their surroundings. One pictograph in particular seemed to all of us the quintessential statement of life among the Anasazi. Life size, painted in ocher on a clean pink cliff, it shows a man standing stiffly with his hand outstretched. Growing from the hand is a tree. In the tree, unmistakable, done with love, is a hummingbird.

Whenever we were within three hundred miles of Fruita, we used to stop. At first we stayed with Doc, later with Dean. If Dean happened to be away, we camped in his orchard. Now, we know, Doc's old motel under the big poplars, fenced with slabs of ripple-marked sandstone, has been cleared away and the land leveled to make the park's picnic ground. Dean's house is now the house of the superintendent. Not even Lurt Knee's Sleeping Rainbow Lodge is open, for the National Park Service has bought up all the in-holdings, as it should have. So we will have to stay outside the park at the Rimrock, spectacularly perched on a stone ridge between the Reef and the Aquarius, with three hundred and sixty degrees of view. Mary, who is tired, is content to sit and look at it. I go in and spend a couple of hours talking to Eugene Blackburn.

Much can be learned from a man like Blackburn, whose memory goes back almost as far as mine, and who has watched all the changes. Though he still has a farm in Bicknell, he has worked for the Park Service for nine years, managing the camp and picnic grounds and running the water treatment plant, and his native attitudes have been tempered by friendly contact with the feds. In his relaxed local voice, without a flat *a* in it—he says "squar" and

Wallace Stegner wished today's visitors could savor Dicey Chesnut's breakfasts. At the Gifford farmhouse, we can come close, with local crafts, jams, and pies for sale in the restored Fruita home, 2008.

"thar" like a character out of the Leatherstocking Tales—he fills me in on what has been happening since the Reef was upgraded to a National Park in 1971.

...All in all, Blackburn gives the Park Service good marks. Pressed, he would probably say it is the best thing that could have happened to large areas of southern Utah. He has only one complaint: he wishes they hadn't torn down the old Chesnut store in Fruita when they started to face-lift the place. Old store like that, with a hand-cranked gas pump out in front, that's practically archaeology, that would interest tourists, especially foreigners. The old days were sort of wrapped up in the Chesnut place. He speaks with feeling; it is clear that his affections are wrapped up in it, too.

I agree. Capitol Reef would be richer if the Chesnut store had been kept. The land is not complete without its human history and associations.

Scenery by itself is pretty sterile. The Chesnuts were living folklore, survivors of the frontier. I would go a good way to have one of Mother Chesnut's breakfasts again, with peaches and cream, hot biscuits, corned elk, and eggs baked in the oven in a muffin tin. A menu like that might be as good a souvenir for a French tourist as a Kodachrome slide of Cohab Canyon, named for the polygamists who hid out here in the 1880s from U.S. marshals trying to serve warrants for unlawful cohabitation.

# 35

## RED ROCKS AND ECCENTRICS (2007)

RENNY RUSSELL

*The Russell brothers, Renny and Terry, grew up in a California family that celebrated art, horses, music, books, photography, and wilderness. Their Aunt Elizabeth was a key influence. When her ailing first husband, Max Lewis, needed clean air, the couple ended up in Fruita, of all places. Elizabeth converted to the LDS Church and continued working as an artist. After Max died, she married her neighbor Dick Sprang (a cartoonist who drew for Batman and Superman comic books). The Russell boys spent the summer of 1958 living with the Sprangs (whose studio is now Ripple Rock Nature Center) and "mingling with Fruita's eccentric inhabitants and merging with the slickrock wilderness surrounding the park."*

*In Renny Russell's 2007 book,* Rock Me on the Water, *from which this excerpt comes, he tells the backstory to* On the Loose, *the classic 1967 Sierra Club book that he and his brother created as young men—a book about their backcountry adventures published just after Terry died in a Green River rapid. Many of us who discovered the thrill of wild country in the 1960s and 1970s remember with tenderness this small book that perfectly matched the times. Fruita and Capitol Reef informed every word of* On the Loose.

In 1958 the small town of Fruita, Utah, was holding on to the last vestiges of a way of life that would soon disappear with the creation of Capitol Reef National Park.... Like the Fremont Indians centuries before, these newcomers relied on the waters of the Fremont River and Sulphur Creek to irrigate their crops.

...Max and Elizabeth bought an orchard along Sulphur Creek in 1954, planning to move there. But cancer snatched Max, and our aunt remarried a gent named Dick Sprang. They built a house and operated a modest fruit business, which they aptly called Ripple Rock Ranch. Phoebe [*Terry and Renny's mother*], Terry, and I spent the summer of 1958 mingling with Fruita's eccentric inhabitants and merging with the slickrock wilderness surrounding the park.

Max left me a bucket of silver dollars for the sole purpose of buying a horse. Joe was no thoroughbred—just a worn-out old cow pony that a local film crew had ridden nearly to death. I rode him along Sulphur Creek—past the old cottonwood tree where the locals used to gather to get their mail before there was a post office; past the schoolhouse built in 1896 that served as a place of worship, as well as a place for dances and the town hall; past petroglyph panels of the Fremont Indians—and then followed the trail along the Fremont River. In a few years, the gorge would be gutted by State Highway 24, which would claw its way through thirty miles of wilderness to the town of Hanksville.

Rimfall, Fremont River canyon, 2011.

I was learning that in the canyon country, the sky could be blue above, but on a distant mesa a thunderstorm could be raging. Floodwater, filthy with desire and urgency, cascaded over the cliff wall and down into the Fremont River. Entire juniper trees and house-size boulders thundered down cliffs. Debris tumbled in a maelstrom of thick, churning, orange water. My horse had seen it all before and was more concerned about the grain back at the barn and getting me off his back, but I had this peculiar desire to put a boat into the foaming torrent and see where it would take me.

I brushed Joe, grained him, and headed back to the house but made the mistake of walking by my uncle's studio windows. Dick was hard at work at his drafting table, and gave me the evil eye. My uncle was tall and solidly built, with large pale-blue eyes like moons that swam in his head and an immense forehead that reflected the sun. Like many of that generation, he wouldn't be caught dead wearing anything but khaki pants and shirts.

Although Dick once said that my Aunt Elizabeth was "a far better and noted artist than I ever was," he was to become a legend drawing *Batman* and *Superman* comics. Dick drew the characters in such a bold and exciting way that he overshadowed the dozens of artists who tried to imitate him. Nobody ever drew a more sinister Riddler or a more expressive and convincing Batman.

During the '30s and '40s, Dick lived in New York City, made connections, was hired on by DC Comics, and for two decades turned out some of the finest comic-book art ever created. At dawn Dick would disappear into his studio, with a cup of coffee to drown the Jim Beam whiskey he'd had the night before. A good workday for Dick was to complete a page of the strip. His talent was too valuable to spend inking, so he sent his pencil work off to ink-master Charles Paris in New York.

I was curious to watch Dick at work, but made my second mistake of approaching him from behind, where he sat trancelike at his drafting table, waiting for Batman to materialize. Dick cut loose a curse, and I ducked a book that whizzed over my head. In his old woody station wagon, my uncle mysteriously came and went. I wouldn't have been surprised to see him flying over the ranch with Lois Lane. A twelve-year-old's imagination can be very active, especially when I watched him clean and oil the Colt .45 he kept on the dresser.

If I couldn't watch Dick work, at least I had his stack of Batman and Superman comic books. In the center of the pile I came upon a *Playboy* magazine, and for the first time, for better or worse, I became familiar with female anatomy.

Our uncle's knowledge of the West was considerable, and his book collection was unparalleled in Wayne County, Utah. After dinner, with whiskey for dessert, he became animated and launched into long discourses on whatever topic popped into his head—Colorado River lore, life in New York City in the '30s (and how it had gone to hell since), or how the Mormons were dangerous. He also made a good case for leaving Capitol Reef as a national monument, saying that if it became a national park it would be ruined.

Dick knew every mile of Glen Canyon intimately—every placer claim, every sandbar, every trail, the name of every cowboy who rode the trails, and even the names of their horses.

Our uncle's eccentric friends who visited included surrealist painter Max Ernst, who fled Europe after World War II and found refuge in Sedona, Arizona. Sprang wrote: "Max would ask me how I ever drew all those little panels so well, and I told him I did it the same way he slammed a paint roller across a canvas and sold it for $75,000. He would roar with laughter and grab me around the shoulders in a big European hug. We knew the score."

Like brilliant artists who suffer from a fire within, Dick could be an aloof nihilist one minute and a gregarious entertainer the next. He moved ghost-like through life—secretive and anonymous. He was a mulish and stubborn man who never forgave or forgot. Still, my difficult uncle's love of rivers and the canyon country of southeast Utah, his knowledge of the West, his talent as an artist and calligrapher, and his aloof independence rubbed off on us. Dick was in the shadows, watching, while we wrote *On the Loose*.

Terry and I befriended one of the hired hands and worked in the orchards loading fruit on a flatbed truck to take to Richfield. I careened through the orchards on a Ferguson tractor, snatching succulent peaches, apples, pears, and apricots.

We had the dismal task of removing mashed apricots from the driveway and quickly learned the difference between a Moorpark and a Sweet Pit apricot. We cranked a machine that turned apricots into slush. We drank it by the gallon, put it on ice cream, and bottled it. At night I dreamed of apricots. On days off, Terry and I hiked down to where Sulphur Creek joined the Fremont River and disappeared into the gorge, where it joined Muddy Creek, becoming the Dirty Devil, eventually combining its silty waters with the Colorado River. The peculiar names of these rivers tugged at my imagination.

Days in ecstasy passed at our favorite swimming hole, where we stretched out on the hot rocks like two chuckwallas, Terry lost in a book and I in a dream. At Capitol Reef Lodge we watched slide shows on geological wonders like Goblin Valley or Glen Canyon. When the lights went out the bats flew in. This was cause for more excitement than the slides generated, especially when bats were caught in a fishnet.

World War II brought electricity, and telephones finally arrived in 1962. A trickling of newcomers began to appear after the war. Although Fruita's population never grew much beyond ten families, the word was out—Fruita had been "discovered." One character in this new wave of outsiders was Dr. Arthur Leroy Inglesby.

Doc was short and stocky, wore thick glasses, chomped on the tail end of a cigar, and wore overalls caked with grease. At the entrance of his home were two signs. One read "Fruita, Elevation 5416." It had arrows that pointed in two directions; one pointed to "Hanksville, 45 miles" and the other to "Torrey, 12 miles." The second sign read "A.L. Inglesby, Rocks and Minerals." Vines and trees obscured his house, constructed from logs and chunks of petrified wood, from the road. Slabs of ripple-marked sandstone were bolted together and formed a fence that enclosed the rich green of his property. He rented out a couple of cabins, mainly to fellow rock hounds.

Inglesby invited us in for lunch. In his musty bedroom a soft tawny light filtered in through the curtain, casting an amber glow on his brass bed, which was nestled among shelves of stones—flowered obsidian, cut spherically to the size of bowling balls; cross sections of dinosaur thighs; bookends made of malachite; and slabs of azurite. Inglesby taught me to cut geodes, showing that hidden inside a rough exterior of a stone can be a sparkling universe of crystals.

One afternoon, while Inglesby was cutting stones and keeping a watchful eye on his rock tumblers, Dean Brimhall stopped to visit; he was another outsider who'd discovered Fruita in the early '50s. Dean spent most of his time rooting around the backcountry with his collapsible aluminum ladders. Some, which were permanently fixed with bolts and spanned great blocks of sandstone, allowed him to reach the high benches overlooking Capitol Reef National Monument. He scoured the alcoves and caves in search of artifacts of the Fremont Indians. Though he was in his seventies, we could barely keep up with his powerful stride. Dean taught that you can't know the desert until you get it into your muscles, and that's just what we did. We explored the monument with him from one end to the other.

In the Moenkopi Formation, dinosaurs once pounded their way through the slime and ooze along the edge of a lagoon or coastal floodplain, when the earth's climate was wet and tropical. The tracks had dried and filled with mud. Dean delicately wedged a crowbar between the layers and tapped with his hammer. After two hundred twenty-five million years in darkness, the tracks of the primitive reptile *Chirotherium* were revealed in the sun, expanding my adolescent concept of time. We loaded Dean's flatbed truck with tracks. By today's standards, that would have been our fourth federal offense—driving off the road, hammering bolts into a cliff, disturbing archaeological sites, and removing dinosaur tracks. Charles Kelly, the only national monument employee, looked the other way.

Like the other characters in Fruita, Charles Kelly left an indelible mark. A printer by trade, he was an amateur archaeologist, a historian, a geologist, and Capitol Reef's first custodian. Kelly belonged to no organization whatsoever, never went out socially, wasn't interested in politics, and hated radios. Rather misanthropic by nature, he harbored a particular contempt for Mormons. He wrote: "I really ought to move to California, but if I did the Mormons would say they ran me out of Utah—so I stay just to spite them."

I'd watch Kelly drive by the ranch in his beat-up pickup on his way to work. Sometimes he would stop for coffee and speak with Dick about the sorry condition of the world, the Mormon problem, the uranium boom, and all those damn people who kept pouring into what the two men considered was their private paradise. He gave me a copy of the book he wrote, *The Outlaw Trail: A History of Butch Cassidy and His Wild Bunch*, which detailed the men who rode through these parts raising hell....

In 1901, it took the Mormon bishop in Torrey ninety minutes to travel to Fruita. Mudslides often closed the road entirely. I watched the highway department straighten out the bends and pave twelve miles of road that snaked down to the monument from Torrey to the west. I walked on the new blacktop highway to the ribbon-cutting ceremony that commemorated its completion. Highway engineers and politicians concluded that the improved road would be the best thing that ever happened to Fruita. I wondered. The locals wondered, too, if the new road possibly was the beginning of the end. In the next decade, their homes in Fruita would be leveled for the new national park. Toward the end of the function, Miss Cora Smith stepped to the microphone with her big ol' hollow-bodied Gibson guitar and belted out an Everly Brothers tune: "Bye bye love; bye bye happiness; hello loneliness; I think I'm a-gonna cry..."

36

## CHARLES KELLY (1987)

GARY TOPPING

*Author, historian, and archivist Gary Topping introduces the founding superintendent of Capitol Reef National Monument, Charles Kelly.*

He was born with a chip on his shoulder. Quick to take offense, slow to forgive, arrogant, self-righteous, Charles Kelly (1889–1971) was a walking argument, a fight waiting for a place to happen. His photographs reveal a man small in stature but compensatingly pugnacious, often with a week's growth of beard and a defiant cigar jutting from his mouth. Savagely antireligious, he lived squarely in the middle of Mormon Country and dared the Mormons to do something about it. As one of the founding fathers of the Utah Ku Klux Klan in the 1920s, Kelly hated even when there was nothing to hate, and confided to his diary that he hoped it would be good for business.

And yet there was good in the man, too. He made friends as fiercely as he made enemies, and though he lost some of them through real or imagined affronts, he remained loyal to most. As a historian, he was a bloodhound on a trail, and if he occasionally followed a false scent, he more often treed his quarry. And he followed his quarry into some of the most remote country Utah had to offer: he took the first automobile across the Salt Lake Desert on the Donner trail, and he floated Glen Canyon several times in the years when it was not generally well known. If he later applauded the flooding of Glen Canyon and rushed into print with the suggestion that the reservoir

Charles and Harriette Kelly, Capitol Reef's first superintendent and his wife, 1950s. Special Collections, J. Willard Marriott Library, University of Utah.

be named "Lake Escalante," he was a formidable defender and student of the backcountry. As the first custodian of Capitol Reef National Monument, he compiled an immense body of interpretive information that fifty years later still serves rangers and visitors alike, and then resigned bitterly in the face of what he considered excessive Park Service bureaucratization.

The first great hate of Kelly's life was his father, a preacher in the mode of Elmer Gantry. Although Kelly danced on his father's grave, he ought to have recalled that he learned two valuable skills during his forced service to his father's profession: music and printing. Always employable through one or the other, Kelly was never without a job, even in the worst of the Great Depression.

Following service in World War I, Kelly settled in Utah with his bride, Harriette Greener. The marriage and the place of residence both lasted the rest of his life. His interest in history came later, as a result of an idle curiosity about the Donner emigrants, but it stuck just as deeply and resulted in a torrent of books and articles, both scholarly and popular, on a variety of subjects including trails, mountain men, Mormons, outlaws, and the deserts and canyons of the Colorado Plateau, all characterized by exhaustive library research as well as interviews and onsite investigations.

# 37

## THE FORGOTTEN PARK

### Charles Kelly (1986)

JONATHAN THOW

*Jonathan Thow wrote his 1986 master's thesis in history at Utah State University on "Capitol Reef: The Forgotten National Park." He majored first in wildlife biology and brought his interest in the outdoors with him to the history department. Thow later became a US Navy judge advocate, an international law expert, and a commanding officer of the Naval Justice School in Newport, Rhode Island.*

In the spring of 1940, at the age of fifty-one, Kelly decided that he had had enough of city life and that it was time to escape to the desert and canyon country of southern Utah. In the years following his new-found interest in history, Kelly had explored every corner of Utah looking for evidence and doing historical research. In the process, he fell in love with the rugged canyons and deserts of southern Utah. In 1940 he sold his interest in the Western Printing Company and started looking for a new home in the desert. He was initially interested in obtaining the superintendent's job at the planned Escalante National Monument, but when the Monument was not established the job opportunity was lost. As Kelly became more and more involved in Western History during the 1930s, he traveled the state exploring many of the rugged areas of southern Utah. Among his many interests was the study of petroglyphs, the engravings left by the early Indian inhabitants of the region. During several of his trips into southern Utah he visited Fruita to view the nearby petroglyphs and to examine the collection of Indian artifacts gathered by E.P. Pectol. As a result of these visits to Wayne Wonderland, he was familiar with the area and had met Doc Inglesby, a retired-dentist-turned-rockhound from Salt Lake City, who had gone to Fruita to retire. On the urging of Inglesby, Kelly sold his house in Salt Lake City in October 1941 and headed for Capitol Reef to make his new home.

When Kelly arrived in Fruita in 1941, his intentions were to buy a small fruit orchard, survive by selling the fruit, live off his savings, and spend his free time writing, painting and exploring the area. However, just weeks later his plans were disrupted by the Japanese attack on Pearl Harbor. The United States' entrance into the War created a demand for all foods, including fresh fruit. As a result, the price of land in Fruita went up almost overnight and prevented Kelly from buying an orchard. When Kelly was unable to purchase any land, he accepted an offer from Inglesby to live in a cabin he had on his property. The Kellys remained with Inglesby for the next eighteen months, surviving off their savings and the meager income gained by the sale of Kelly's articles to the *Desert Magazine*. During his first months in Fruita he had the opportunity to meet with and to become friends with Paul Franke, the superintendent at Zion National Park. In the spring of 1943 his friendship with Franke gave Kelly the opportunity to work for the National Park Service.

When Capitol Reef National Monument was established in 1937 it had been placed under the administrative jurisdiction of Zion National Park until it could be activated. In the early 1940s the National Park Service bought the so-called Chesnut property in Fruita to gain the water rights they needed for later development of the Monument. In March 1943, Franke offered Kelly the position as Monument custodian in exchange for the use of the house and the orchards that came with the Chesnut property. The position was without pay, but with the house, orchards and the income from Kelly's writing, he and his wife could survive until they were able to buy their own property. However, the end of the War did not bring the land prices down and Kelly ended up remaining in his voluntary position for the rest of the decade. He continued to look for a way out of Fruita, but "inflated prices [made] it impossible to buy land or property anywhere," so he was "stuck" there a dour and often surly man. By 1949 Kelly had decided that with no development there was no future for the area as a National Monument and that it was time for him to move on to other things. In the next few months, however, events took place that resulted in Kelly remaining at Capitol Reef National Monument.

In August 1949, Kelly took the Civil Service examination to qualify for the soon-to-be created position of superintendent at Capitol Reef. He passed the exam and, after being initially turned down because of his age, was given the position of full-time Ranger when the Monument finally went on active status on May 1, 1950. In 1953, his position was upgraded to superintendent and he was given a small raise in pay. Kelly remained as the superintendent of Capitol Reef National Monument until February 1959, when his failing health and eyesight forced him to retire shortly after his seventieth birthday. The Kelly years were marked by conflict over uranium mining…and the development of the Monument,…but in addition to these important events, Kelly left behind his legacy at Capitol Reef when he retired to Salt Lake City in 1959.

While Charles Kelly truly loved Capitol Reef, in many ways it is unfortunate that he became the representative of the Park Service in Wayne County in 1943. When Kelly arrived he found a county that was overwhelmingly dominated by the Mormon Church. While the figures to determine the exact number of Mormons in the county are not available, it is safe to say of the 2,400 people listed on the 1940 Census, that more than 95 percent of them were members of the Mormon Church. [*Change and growth come slowly. In 2017, 2,719 people lived in Wayne County. In 2014, 74 percent of them were LDS.*] The communities near the Monument were very small. Most of the county's residents were still living in much the same manner that their ancestors who settled the region had done and they shared many negative attitudes toward non-Mormons and outsiders. The Park Service, by installing a man like Charles Kelly as its sole representative in the area for almost sixteen years, showed a total lack of understanding for the residents of the area. It is not difficult to imagine how a man who hated most things associated with religion and who had made a practice of baiting Mormons since his 1919 arrival in Utah, was received by the residents of the area. It also takes very little imagination to get a sense of how Kelly responded to the people of the region.

Kelly fought a running battle with the Mormons of Wayne County for the sixteen years he represented the Park Service at Capitol Reef. At one point, he claimed "the Mormons of Wayne County tried to replace me with a Mormon, but I told them off and pulled a fast one... so they won't try that again." The details of this particular incident were not recorded by Kelly, but his attitude toward the people involved is very apparent. Kelly neither liked nor trusted the Mormons living around him. In terse terms filled with innuendo, he reported to his superiors in July 1955 that

> the banker at Loa, who is also in the cattle business, left a dead calf near the campground. He was asked to remove it.... The bishop of Torrey dumped a large quantity of beer cans on the side of the road. He was asked to remove them.... The banker at Loa and the bishop at Torrey are no longer friends of the superintendent.

It is difficult to imagine that Kelly was ever very friendly with the banker and the bishop, but this incident seems to have ended what cooperation had existed between them. Kelly also had continuing trouble with local youths coming on to the Monument and using it as a place for their parties. At one point he "had to get tough with two groups of young Mormon hoodlums during the summer" of 1954, and he decided, as a result of this incident, that he would "no longer attempt to be friendly with the local people." Kelly also had trouble during the 1950s with the people still living in Fruita allowing their cattle to graze freely on the Monument lands. Each time this happened, Kelly made the owners remove their cattle and occasionally pay for damage they had caused to the Monument. This action made Kelly the focus of the growing hostility against the grazing ban that had come with the creation of the Monument, but that had never been enforced until Kelly's arrival.

While it is very likely that the relationship between the Park Service and the residents of the county would have deteriorated over time, there is no question that the Kelly Era at the Monument hastened this process.

The Waterpocket Fold, Swap Mesa, and the Henrys from the Burr Trail switchbacks, 2016.

# PART VII

## NATIONAL VISION

### Crafting Wayne Wonderland

With its longtime lack of wide renown, Capitol Reef merited historian Jonathan Thow's label "The Forgotten Park." In 2017, Zion National Park received four times as many visitors as Capitol Reef, Bryce Canyon more than twice as many, and Arches National Park a third again as many visits. But Capitol Reef is not the least-visited of Utah's five national parks. Canyonlands holds on to that quiet distinction (in 2017, 742,000 visitors vs. Capitol Reef's 1.15 million).

With the Utah Travel Council's 2013 advertising blitz for the "Mighty Five" national parks and a surge of interest during the 2016 centennial of the National Park Service, these numbers will only go up—as they have at Capitol Reef, with visitation doubling since 2013.

And yet a core of rural Utahans continue to resent all special designations on federal land, though they depend economically on the waves of tourists and newcomers that protected lands bring to gateway communities. Capitol Reef is at the center of this history.

Historians Jared Farmer, Jonathan Thow, and Bradford Frye take turns introducing us to these stories of controversy and conflict. The narrative begins in the 1920s, with passionate visionaries in Wayne County who hoped to boost local business with a Wayne Wonderland National Park. A huge proposed Escalante National Monument delayed their dream, and a small Capitol Reef National Monument that lacked services for decades soured their enthusiasm.

The park has remained at the center of national confrontations between those who use land as commodity, who see Nature in need of "finishing," and those who treasure land for its wild self, who view humans as just one participant in communities of the living and nonliving.

In the 1950s, Congress opened Capitol Reef to uranium mining over strenuous objections from Superintendent Charles Kelly. The surge in environmentalism at the time of the first Earth Day in 1970 created an opportunity for monument expansion during the tenure of conservation-minded Secretary of the Interior Stewart Udall. In the last hours of his presidency, Lyndon Johnson used the Antiquities Act and signed a proclamation to increase the monument area by six times. Lurt Knee gives us his insider-outsider look at this decision.

Reaction in southern Utah to the monument's enlargement was predictably feverish, with cattlemen especially concerned about the future of their grazing permits, which the planning process had not grappled with effectively. Republicans in Utah's congressional delegation supported these fervent advocates of multiple use. But by the time Utah conservationists George and Gene Hatch helped Utah senator Ted Moss push through national park designation two years after the monument expansion, in 1971, the controversy had begun to dim. Not completely, however. The final park boundary included Halls Creek and its narrows at the southernmost reach of the Fold, and planners soon closed that spectacular route to motorized vehicles. Not everyone rejoiced.

The Burr Trail, the park backway that climaxes at a precipitous series of switchbacks climbing the cliffs of the Waterpocket Fold, sparked more controversy in the 1970s and 1980s. Should this adventurous backcountry passage remain unpaved? Who owns the right-of-way? Who makes these decisions?

38

# FIX THE ROADS, THE TOURISTS ARE COMING (1999)

JARED FARMER

*Historian Jared Farmer grew up in Utah and teaches at SUNY, Stony Brook. Like so many who have written about Capitol Reef, his interests span many disciplines, including the overlapping historical dimensions of landscape, environment, science, culture, and religion. His books about western environmental history include* Glen Canyon Dammed: Inventing Lake Powell and the Canyon Country *(1999), from which the two excerpts in this section come.*

Just as San Juan County had Zeke [*Johnson, the local booster of Natural Bridges National Monument*]—someone who selflessly and tirelessly worked to promote the beauties of his home region—Wayne County had Joseph Hickman and E. P. Pectol. Taking inspiration from Zion National Park, these brothers-in-law formed a two-man booster club in 1921. Hickman and Pectol believed that the red cliffs of "Wayne Wonderland" were just as beautiful as Zion and deserved to be recognized. In 1924 Hickman won a seat in the legislature, and he took his cause to Salt Lake City. The result: the Board of State Park Commissioners was established with the purpose of identifying and creating parks. Hickman clearly meant Wayne Wonderland to be first on the board's list. In July 1925, he and other county leaders staged a local celebration for "Wayne Wonderland State Park." There was a rodeo, a dance, and a Sabbath-day assembly in the town of Fruita (now the location of the Capitol Reef National Park visitor center). Many speakers, including the governor, looked forward to the day when the state, and ultimately the National Park Service,

Fruita schoolkids welcome Governor George Dern to Capitol Reef, 1925. Courtesy Capitol Reef Natural History Association.

would assume responsibility for the scenic country adjacent to Fruita. With federal designation would come good roads, something the state couldn't easily pay for on its own.

Sadly, Hickman died in a boating accident just two days after the festivities. Despite the loss, Pectol carried on. A natural leader—for fifteen years he served as the local Mormon bishop—he helped organize an umbrella group to address southern Utah's persistent economic problems, made even worse by the Depression. The charter meeting of the Associated Civic Clubs of Southern Utah (ACCSU) took place in Richfield in 1930. Ten counties were represented. "The purpose of the associated clubs," reported the local paper, "will be to foster, encourage and provide by every honorable means the growth and development of the communities of southern Utah . . . to give publicity to the scenic and commercial resources of southern Utah, and to exploit systematically the products of southern Utah for exporting and consumption."

The ACCSU considered road improvement one of its primary goals and supported national park status for Wayne Wonderland in large part because it promised to stimulate interest in an east-west highway linking Mesa Verde National Park to Zion, Bryce, and Grand Canyon. The result, proponents believed, would be a "golden circle" of accessible scenic areas in the Four Corners states. Tourism seemed guaranteed. It was time, the members of the ACCSU believed, to make this beautiful, harsh land pay. So far, despite the best efforts of the Mormon pioneers and their descendants, the canyonlands had yielded little. Maybe this was God's Country, but from an economic perspective it seemed like The Land that God Forgot. By emphasizing the former, however, tourism boosters believed they could help alleviate the latter. This spirit of entrepreneurship was well expressed by J.E. Broaddus, a Salt Lake photographer who, working with Ephraim Pectol, produced a promotional slide show about Wayne Wonderland. "Utah people have failed to realize," he said, "that they have a great industry in the promotion of their scenic attractions and have also failed to see that they have a greater variety of scenery than any state in the Union."

The National Park Service held a similar opinion about the people of Utah. Unfortunately, federal officials, like most outsiders, rarely acknowledged the efforts of local boosters like Broaddus and Pectol. They preferred, it seemed, to categorically belittle locals for blindness. With high paternalism, Robert Sterling Yard, an early publicist for the Park Service, once wrote, "People had to 'come in' before [Bryce Canyon] was appreciated. And so it will be with...the entire great region of the High Plateaus, where people have not yet come in. Other national parks no doubt there await discovery."

Several years later, the agency found a big one: Glen Canyon of the Colorado River. Following the bold, sometimes imperious leadership of Interior Secretary Harold Ickes, the Park Service staked its claim to the canyon in 1936, the same year Robert Marshall [*prominent forester and founder of the Wilderness Society*] published his roadless area inventory [*recommending protection for millions of acres of federal lands*]. The proposed Escalante National Monument would have covered 6,968 square miles, twice the size of Yellowstone National Park, or about 8 percent of Utah. "Naturally and logically," wrote a Park Service official, "the question arises as to why this great area of outstanding scenic resources has not become known to the public generally. The simple and direct answer may be synthesized in a simple word—inaccessibility." Indeed, the boundaries of the megamonument generally agreed with Marshall's Colorado River roadless section. The Park Service surely would have broken the roadlessness with many scenic drives, but, as one supporter wrote, "Even after roads have been built into this magnificent section, it will be generations before its hidden recesses have been fully explored and therein lies its fascination."

This unsurveyed region contained one "town" (Hite, population ten—give or take nine), and an estimated 27,000 cattle and 218,000 sheep. The owners of these animals lived in outlying settlements such as Moab, Monticello, Loa, and Escalante. In the 1930s, grazing constituted the single great economic use of the monument area, especially before the Taylor Grazing Act of 1934 (which initiated federal regulation under the Grazing Service, predecessor of the BLM). Under normal Park Service regulations, however, all grazing would have been eliminated. Disgracefully, only after its proposal went public did the Park Service, under pressure from the state of Utah, seek out the opinions of local ranchers. Eighty-seven appeared at a public hearing in Price, Utah, in June 1936. David Madsen, a low-level Park Service employee and a lifelong Utahn, had the unenviable task of explaining the motivations of his superiors in Washington. In the meeting he tried to sympathize with the ranchers and expressed his hope that provisions could be made for the continuation of grazing rights. However, he suggested that the highest economic use of the land was not ranching but recreation.

Charles Redd, one of the state's most respected stockmen, stood up and made the opening reply. The people of southern Utah "do, and have for a long time, recognized the value and importance, the unusualness of the scenery we have here," he said. However, "the exploitation of these natural wonders is not inconsistent with the full and free use of the range by livestock." To the assembled ranchers, an attack on their labor was also an attack on their land. Where some people saw only beautiful wild country, Redd and his peers saw wild, beautiful country made better by three generations of Mormons. Though they understood that the American people technically owned the land, they didn't necessarily believe the public deserved it. "Now, I am willing for the people who wish recreation to have rights

The dedication and celebration for the new Capitol Reef National Monument at "Singing Rock" in Grand Wash attracted 2,500 people on September 25, 1937. Photograph by A. L. Inglesby. Courtesy Utah State Historical Society.

and privileges," said W.A. Guymon of Huntington, "but I don't think that it is fair for them to tell me to move off simply because my land has some kind of an attractive object that the people think should be given to them. Let them pay the price." Better yet, let them get their money's worth at Utah's existing recreation areas. In a resolution drafted after the meeting, the ranchers called attention to Natural Bridges National Monument (1908) and Arches National Monument (1929). "The Park Service to date has expended nothing to make them accessible to the traveling public; and we strongly maintain that the monuments already created should be developed before vast additional areas are withdrawn for recreational purposes to the expense of our basic industry."

Reacting to opposition from ranchers and politicians in Utah, the Park Service trimmed its proposed monument to 2,450 square miles in 1938. Still skeptical, the state of Utah pushed for ironclad safeguards on future grazing, mining, and dam building. The Park Service resisted, and the proposal soon degenerated into bickering and miscommunication between the state and federal governments. Eventually, Escalante National Monument lost momentum within the Roosevelt administration and died. Thus Glen Canyon remained the same—without public roads or recognition or protection.

Forgotten amidst the controversy, however, was the creation of a much, much smaller national monument in Wayne County, a monument closer to the ideals of rural Utahns. The big break occurred in 1933, after Ephraim Pectol attained Joseph Hickman's old seat in the legislature. From there he shepherded a unanimous resolution to Congress urging the establishment of a Wayne Wonderland park and a regional highway connecting it to the "chain of natural wonders" in the region. Pectol also secured the support of Utah's congressional delegation. In response, the National Park Service authorized a feasibility study, the first of many bureaucratic hurdles. Pectol and the ACCSU continued to lobby and worked to allay the concerns of local ranchers, who didn't share their enthusiasm for tourism. Ultimately, in 1937, President Franklin D. Roosevelt created Capitol Reef National Monument by executive order. (Many in Utah disliked the monument's name; Congressman Abe Murdock tried unsuccessfully to reinstate Wayne Wonderland.) Said Governor Henry H. Blood at the dedication, "Capitol Reef is but a beginning, roads must follow—and with roads must come development and new areas to be opened up."

It took more waiting, however. In the words of one historian [*Jonathan Thow*], "Roads were very slow in coming, few trails were built, no campgrounds were set up, no administrative structure was put in place, and no scenic highway was constructed connecting Capitol Reef with the other monuments and parks in southern Utah." The monument didn't even have a budget until 1950. As such, it contributed little to Wayne County's economy. Understandably, locals felt betrayed by the Park Service. For decades, the national monument sat in hibernation, to the chagrin of tourism boosters.

# 39

## THE FORGOTTEN PARK

### Uranium Mining (1986)

JONATHAN THOW

*This excerpt comes from Jonathan Thow's master's thesis in history at Utah State University.*

While the early construction of highways into Wayne County and the Mission 66 development of the Monument were progressing during the 1950s, another crisis between the Park Service and the local residents was growing out of the uranium boom in the years following the Second World War. The region around Capitol Reef is not rich in natural resources. It does not contain any of the valuable metals that are found in other areas around the West. However, this barren land of rock and sand does contain trace elements of uranium. After World War II, the United States was thrust into the Atomic Age and the era of the Cold War. Uranium, a heavy radioactive metal used today in atomic weapons and as fuel for nuclear power plants, became a vital component in the development of our nuclear arsenal. The paranoia of the Cold War led the United States into an arms rush and pushed it to build bigger and more destructive weapons, creating a new market for uranium and causing something of a uranium craze to sweep the nation. People from all over the United States headed West with high hopes of making an overnight fortune and returning home to enjoy their new found wealth. While proportionally the number of people involved was certainly smaller, the attitudes, ambitions and fate of the new prospectors can be closely compared to those who headed for California during the 1849 gold rush. As happens during any mining boom, a small number became wealthy. However, the overwhelming majority suffered great hardships and finally quit the land with little to show for their efforts.

A great deal of the uranium speculation was focused on the Four Corners region where Arizona, Utah, New Mexico and Colorado share a common border. Prospectors fanned out across southern Utah as far north as the area around Capitol Reef National Monument. However, because of its monument status, no prospecting was allowed within its boundaries. The blockage not only angered prospective miners, but also upset residents of the county who hoped the uranium boom would revive their struggling economy... the anticipated tourist trade to the Monument had not yet developed in the early 1950s and it was beginning to appear that it might never grow. At a time when the county's population was declining and its eco-

US Geological Survey geologist R. Q. Lewis at the Oyler Mine, Grand Wash, 1952. Photograph by E. N. Hinrichs. Department of the Interior/US Geological Survey.

nomic future bleak, it was difficult to understand why the resources of the Monument could not be exploited. The resulting fight over mining rights in Capitol Reef became another major point of contention between Wayne County and the Park Service. Indeed, the 1950s turned into a running battle between Charles Kelly and the various groups attempting to prospect and mine inside the Monument.

While Kelly's fight was ultimately against several broad forces, including the Atomic Energy Commission (AEC) and the Park Service itself, there was one claim within the Monument that caused him more trouble than any other. This was the so-called Oyler Tunnel case which dragged on for several years and came to involve a number of local people. By the time it was finally settled, it had further fueled the fires of controversy between the Park Service and Wayne County....

In February 1951, the AEC wrote the Department of Interior expressing the belief that the area around the Oyler Tunnel [*the first uranium mining claim in the park, at the mouth of Grand Wash on the Scenic Drive*] might contain large enough deposits to justify working the site. Under normal conditions the reply from the Park Service would have been no; however, the advent of the Cold War and the rush to build a nuclear force made people very responsive to the will of the AEC. Throughout the decade, few effective challenges were raised to the AEC's policy and it got virtually anything it needed. After several months of negotiations, the AEC was granted a seven year limited use permit by the Park Service, for the prospecting and the removal of uranium from Capitol Reef National Monument.

Charles Kelly, the superintendent of Capitol Reef during the 1950s, had a little different view of how the Monument came to be opened to mining. It was his opinion that the AEC became interested in the area around Capitol Reef after Christensen [*Willard Christensen, a local miner with a claim on the Oyler Tunnel*], whom Kelly called a "lunatic," wrote a letter to his congresswoman complaining about his mining claim at the mouth of Grand Wash being blocked by the Department of Interior and the courts. Kelly claimed that Christensen

> set down and wrote a letter to [Utah Congresswoman] Reva Beck Bosone in Washington and told her there was millions of dollars

> worth of uranium tied up in the monument and the Park Service wouldn't let him dig it out. The nation was just starved to death for uranium and was liable to be blown to hell by Russia if we did not get that uranium out of the monument. Now Christensen is an ignoramus that doesn't know how to spell and he doesn't know any grammar and it was a hard job to try and figure out what he had to say in that letter, written on cheap paper with a pencil. So she read it and started to cry her eyes out because this poor old prospector had been denied the right to dig out the uranium, and she took it up with the Park Service, and on account of all the excitement of the uranium prospecting around the country, the senators and the representatives back there got all worked up about all the millions of dollars of uranium that was tied up in Capitol Reef National Monument. So without any examination of any kind they arbitrarily opened it up to mining under these special permits that they issued.

Because of all the trouble Kelly had experienced keeping Christensen and others like him from mining in the Monument, he probably overstated what had happened when Christensen wrote Washington, but regardless of how it happened, the result was the same. The Oyler Tunnel, probably with help from Christensen and other local miners, attracted the attention of the right people in the nation's capital. On February 19, 1952, an agreement was reached that opened one of America's national monuments to prospecting and to the eventual mining of uranium.

By the time the Monument was actually opened to mining in 1952, the area around the boundaries had already been checked and rechecked by prospectors, and thousands of claims filed. None, however, proved to be a commercial success. Despite the lack of findings beyond the Monument perimeter, hundreds of people still applied for permits for uranium prospecting within it.... One group of miners went so far as to rent an airplane and spend two days flying over the area with a geiger counter. Even this failed to show evidence of commercial uranium deposits within the Monument.

...Despite hundreds of prospectors combing the Monument, only five claims had actually been registered and no mining permits had been issued within the Monument by February of 1953. In a moment of optimism in February, Kelly declared that the uranium boom was over, but he was proven to be badly mistaken in the next few months.

...By the end of May, the AEC had granted mining contracts to Lurton J. Knee, the Kuhn Brothers, Robert Dunsmore and the La Florecita Mining Company to remove uranium ore from the Monument. With this action, the mining of uranium started in earnest and with it came the disruption of the landscape, as Kelly had feared.

Distressed because of all the mining activity, Kelly fired a letter off to the AEC in May protesting the mining contracts. He was angry because the fragile landscape of the Monument was being destroyed by roads, tunnels, dumps and other trash. Kelly felt because no ore was being shipped, that the contracts had been issued without the AEC having properly checked out the claim sites. In his letter, he complained "if the practice is continued the Monument will be pockmarked with abandoned tunnels and dumps, with no ore recovered." Kelly's letter reached the right people within the AEC. In July they sent Ray Lindbloom to investigate Kelly's complaints. The crusty old superintendent took Lindbloom on a tour of the Monument, showing him all the damage that had been done. Lindbloom was convinced, agreeing that under the conditions no mining contracts should have been issued. After studying the entire matter carefully, Lindbloom concluded that contracts would not be issued in the future without specific investigation of the sites. However, he allowed those issued to remain in effect....

As winter turned to spring at Capitol Reef in 1954, the area braced itself for yet another influx of people looking for an instant fortune from uranium. Beginning in February the rush focused on the region south of the Monument in Garfield County. Kelly wrote in his monthly report for February that the rush had drawn almost everybody out of Wayne County and that people had closed up their shops and dropped everything to head for the new El Dorado. Indeed, the rush drew so many people out of the county, that some Mormon bishops even packed up and headed south to join their congregations. The uranium fever lasted for three months, ending only after several thousand new claims were filed by eager prospectors. At its highest pitch the rush of 1954 reached such a frenzy that prospectors were jumping other claims in search of the one that would make them wealthy. Kelly wrote that

> prospectors' jeeps passing through the Monument have been averaging 40 a day.... They all drive with the throttle wide open in order to beat the other fellow to those million dollar claims. Dust in the checking station is an inch deep every night. The gold rush of '49 was never like this.

Despite all the activity, no major strike was made in the area south of Capitol Reef....

On October 13th Kelly wrote to the regional director of the Park Service in Santa Fe recommending that the agreement between the Park Service and the AEC be canceled. He felt in view of the fact that the prospecting of the previous eighteen months had turned up nothing of worth, that the Monument should be returned to its intended condition. Kelly suggested that the special permits for prospecting not be issued after February 20, 1955. At the time, in a short paper outlining the history of the uranium mining in the Monument, Kelly wrote that

> hundreds of prospectors have searched every inch of ground and spent thousands of dollars blasting out tunnels and prospect holes. Every foot of ground has been staked and restaked. And yet not one pound of uranium has been extracted and sold, for the simple reason that it does not exist except in one very limited locality, and even that is not extensive enough to be worth mining.

...In early May, "after lengthy and sometimes heated correspondence," the AEC agreed to stop issuing permits for prospecting in the Monument on May 17, 1955. The contracts that had been previously issued for mining in the Monument would be valid until May 1956. Unless one of the active mining sites hit a major uranium vein, all mining in the Monument would be suspended no later than May 1956....

The peak of the mining activity at Capitol Reef lasted from 1952 to 1956. During that period, uranium yielded very little to the federal government, which Kelly alleged was paid only about thirteen dollars on its share of the uranium produced within the Monument. When it is considered that the government received 10 percent of the income in royalties, it is clear that the area did not contain vast amounts of uranium. During the four year period of mining activity, Kelly fought a running battle to keep the Monument from the ravages of mining. To a large degree he was successful, damage to the land was relatively minor and it healed within a few years. However, the damage to the relationship between the Park Service and the miners, many of whom were from the area, was very slow in healing.... The people of the county did not really understand the purpose of Capitol Reef National Monument.... In their minds, the Monument existed for their economic benefit, not for the protection of a unique scenic and geological area.... When the expected tourist boom did not materialize during the early 1950s, local interests wanted to exploit the land in other ways for economic gain. It was a natural reaction, given the economic condition of the region, but it demonstrated the misconception about the Monument that many people in the region held. The lack of understanding, however, was not a one-sided problem.... The problems that grew out of the Kelly Era and flared up over uranium mining in the 1950s, were the beginning of a much larger problem that continued to haunt the relationship between the Park Service and Wayne County.

# 40

## FROM BARRIER TO CROSSROADS

### The Genesis of Expansion, 1968 (1998)

BRADFORD FRYE

*Brad Frye based his administrative history of the park on exhaustive archival research and his own collection of interviews and oral histories. He earned a master's degree at Eastern Washington University with this work. The National Park Service published his history in 1998, and it's available in full online.*

The first that Capitol Reef Superintendent [*Robert*] Heyder heard about proposals to expand his monument was at the dedication of Carl Hayden Visitor Center overlooking the Glen Canyon Dam on September 26, 1968.

Heyder was only 37 years old, but had been in the National Park Service since his youth at Yosemite and Grand Canyon National Parks. Most recently, he had served as management assistant at Wilson's Creek National Battlefield in Missouri. His long career would see Heyder assume the superintendency of Bryce Canyon, Zion, and Mesa Verde National Parks before his retirement in 1993. He had been at Capitol Reef National Monument for only about a year when he was invited to hear Secretary [*of the Interior Stewart*] Udall speak at the Hayden Visitor Center dedication. During the pre-ceremony gatherings, Regional Director Frank Kowski asked Heyder if he were aware that the park service was considering enlarging Capitol Reef National Monument.

When Heyder said he knew nothing about this, Kowski explained that under consideration was a separate monument unit encompassing a portion of the Waterpocket south of Oak Creek to just above the Burr Trail. The stock driveway, water diversion, and irrigation dam in the drainage may have been why Bates Wilson [*superintendent of Arches and Canyonlands National Parks and canyon country park advocate*] and/or other field office personnel decided to exclude Oak Creek itself, and not connect this new unit directly to the southern boundary of the monument.

When Heyder heard about this initial expansion idea from Kowski, he immediately pointed out that such a small additional unit was not enough. Heyder argued that any new boundary must be extended to ensure protection of existing monument resources. Heyder also recommended that the entire Waterpocket Fold, from the Fishlake National Forest boundary in the north to the Glen Canyon National Recreation Area boundary in the south, be included in the expansion. Heyder later recalled that the regional director seemed surprised by these ideas, and promised to talk to him further after the dedication ceremony.

After the dedication speeches by Udall and others, Kowski introduced Heyder to Secretary Udall. This was the first interior secretary Heyder had ever met and the meeting would prove to be particularly memorable. According to Heyder, Udall asked him, "Oh, what do you think about this idea of widening [the monument]?" Heyder responded that he didn't think the expansion was being done right. When the secretary asked what Heyder would do, the superintendent spoke up:

> [Y]ou ought to go west of the main body [of the existing monument] and east a bit and take in Cathedral Valley up there and come off the [Thousand Lake] mountain and come all the way to the forest, you know, and take that whole thing in and then go all the way down to the recreation area.

Udall responded favorably to Heyder's suggestions. Heyder recalled, "He said, 'You get with Bates the next couple of days and come up with something.'" Thereafter, the specific proposal to expand Capitol Reef would be largely in the hands of Superintendent Heyder.

Morrell cabin, Cathedral Valley, 1975.

The day after the dedication, Heyder drove over to Moab and consulted with Bates Wilson. There he saw a map of the original proposed boundaries, and they discussed Heyder's ideas. Wilson then directed Heyder to "go back and pull the whole thing together and send it in and send me a copy."

When Heyder returned to Capitol Reef, he took Chief Ranger Bert Speed into his confidence, and the two of them worked "the better part of three nights" poring over the maps and coming up with various alternatives. The first closely approximated the final January 1969 proclamation boundaries. Other possibilities, which considered adding the Henry Mountains or the Circle Cliffs, were not submitted because Heyder feared the acreage would be too big, too controversial. Later, a separate proposed boundary drawn by agency officials included the Circle Cliffs. The maps and descriptions were sent on to the regional office and from there directly to Secretary Udall. All of this work was accomplished secretly. In fact, the text of the proposal was typed by Heyder's wife, since his secretary was Afton Taylor, wife of Wayne County rancher Don Taylor.

... In a memorandum to the president dated December 5, Udall listed the national monument and wildlife refuge possibilities. In order, they were a 2 million-acre addition to the southern half of Mt. McKinley National Monument in Alaska; establishment of a 3.6 million-acre Arctic Circle National Monument, Alaska; a new 26,000-acre Marble Canyon National Monument, Arizona; a new, 911,000-acre Sonoran Desert National Monument, Arizona; a 94,000-acre addition to Alaska's Katmai National Monument; a 49,000-acre addition to Arches in Utah; and the 215,000-acre proposed addition to Capitol Reef National Monument in Utah. Also included were two additions to Alaska wildlife refuges totaling an additional one million acres.

It is unknown whether the list was created in order of preference or potential controversy but, in any case, Capitol Reef was far less important in the ensuing debate than the huge proposals for Alaska, or than a Sonoran Desert Monument that would encompass a military firing range in southern Arizona.

During the closing months of Johnson's administration, the president increasingly surrounded himself with a circle of advisors. One of his closest advisors, Special Counsel W. DeVier Pierson, would become the key liaison between Udall and the president in the weeks to follow. In his analysis, which accompanied Udall's December 5 memorandum, Pierson initially supported Udall's proposal, telling Johnson, "It would be the last opportunity to cap off your exceptional record of additions in the park system."

On December 11, Udall presented his case, complete with slides, maps, and graphs, to the president and the first lady. The presentation was followed by cabinet members and aides grilling Udall on the proposed areas and on the general consequences of the president issuing such controversial proclamations during his last days in office. After the meeting, Udall believed he had persuaded Johnson to sign, once a few minor legal questions were settled. The secretary was convinced the whole package would be signed by mid-December, as Johnson's "parting Christmas gift to the American people."

Unbeknownst to Udall, Pierson was now questioning the political implications of adding such a tremendous amount of acreage to the national park system without congressional approval. On December 12, Pierson wrote the president to point out that, while the Utah and Arizona monuments engendered little controversy, the Alaska proposals were particularly sensitive. ...The secretary later recalled:

> [Johnson] was much too concerned about congressional reaction, because I couldn't clear this all the way through Congress. After all, after January 20, his relations with Congress are not important. He didn't have any legislation to get through. The question was had he done what he thought was right for history and right for the country in terms of a final conservation achievement.

Clearly, Udall thought of this entire issue as the deciding legacy he and Johnson would leave to conservation history. Johnson and Pierson, on the other hand, were more concerned with a possible legacy of controversy during the president's last days in office.

The deadline for approving the wildlife refuges in Alaska was 30 days before Nixon was sworn into office on January 20. While Udall could have authorized the wildlife refuges himself, he really wanted his entire package signed by Johnson in those last days before Christmas. Johnson, however, postponed signing any proclamations because he was dissatisfied with Udall's minimal congressional checks in mid-December. Further delay occurred when Johnson became ill and was hospitalized for several days, and then spent Christmas recuperating at his home in Texas. When Udall pressed about the monument approval, Pierson replied that Johnson would not decide until he personally contacted congressional leaders in January.

Throughout early January, Udall continued to push for approval and Johnson kept delaying final action. On January 14, Udall delivered a breakdown of the extensive use of the Antiquities Act by previous presidents during their last weeks in office. This was attached to a memorandum to Johnson stating that everything would soon be in order and that the president could proceed with the signing the next day. An accompanying memorandum from Pierson advised continued caution:

> I still have reservations as to the desirability of taking this action during the last week of your Administration. However, some of the

> proposed areas are very exciting. Consequently, you may wish to examine them on a case-by-case basis and act on some while deleting others.

Pierson included an annotated list of the proposed national monument additions for the president's review. Capitol Reef National Monument was lumped with Arches and Marble Canyon national monuments and was "justified on the basis of unique geological or scientific qualities." Pierson had no problem with these acquisitions, provided the congressional delegations went along. The small Katmai addition posed little problem, but the proposed Sonoran Desert National Monument and the Gates of the Arctic and Mt. McKinley additions were generally seen as too large and too controversial.

Johnson wrote "OK" next to Capitol Reef, Arches, and Marble Canyon; "maybe" next to Katmai and Mount McKinley; and nothing next to the other two. The president also noted that he wanted further congressional checks "at once in depth." Thus, it seemed that Johnson was moving toward a compromise to include only the smaller additions—once congressional leaders were consulted. Capitol Reef's 215,000-acre addition would increase the monument's size by six times, yet this paled in comparison to the almost 6.5 million-acre proposals in Alaska and southern Arizona.

It is unlikely that Udall was aware of this new course toward down-sizing his original proposal. On Friday, January 17, only three days before the inauguration, Udall reported to Johnson that he had discussed the idea with the Utah delegation, which had "a surprisingly good reaction." He wrote, "Even Sen. Bennett, who fought the Canyonlands National Park, favors our proposal...."

At the same time...the interior secretary found himself trying to keep the lid on press releases detailing the president's signing of all 7 million acres. These releases were apparently written specifically to answer questions raised by the president's State of the Union address on the previous Tuesday, when Johnson had ad-libbed that he was not yet finished with his conservation effort. This veiled reference forced Udall to field inquiries from the press and concerns from congressmen over exactly what Johnson intended.

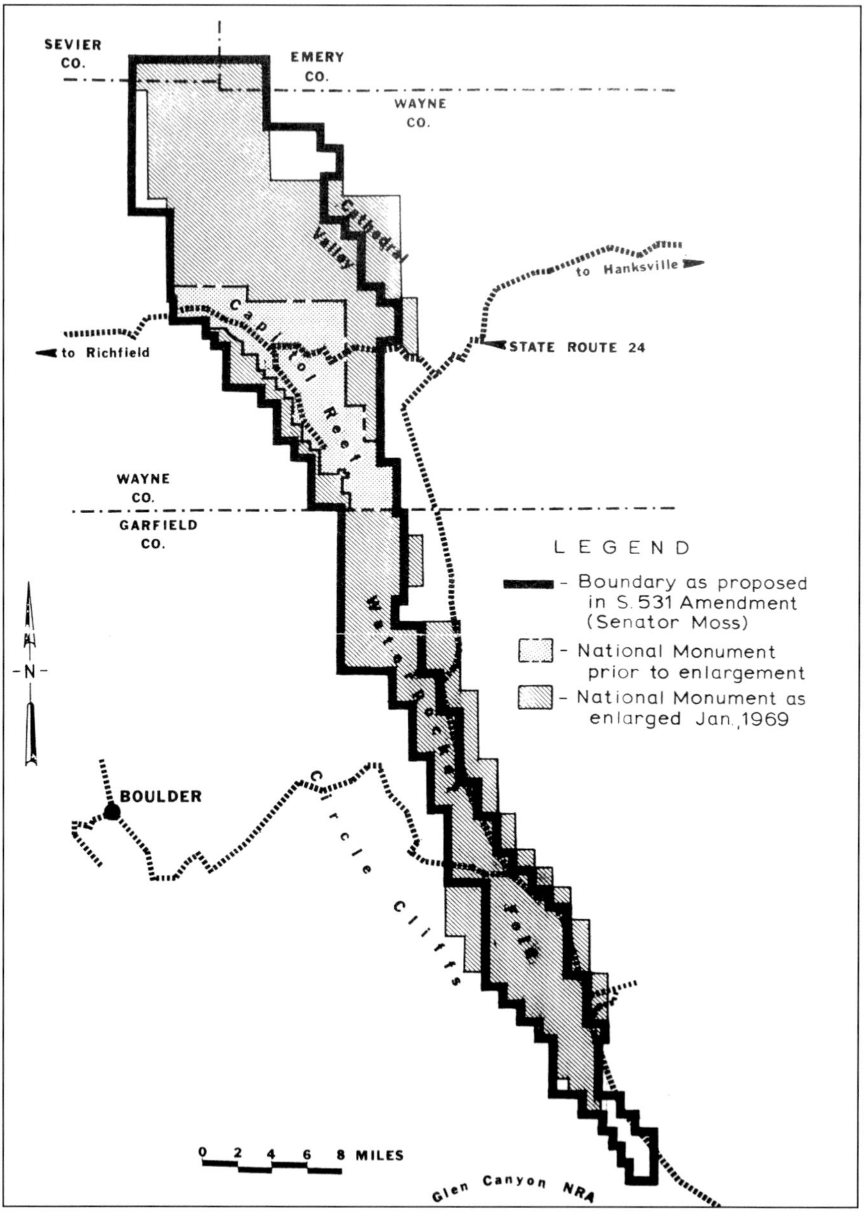


Capitol Reef boundaries through time.

Udall and his director of information, Charles Boatner, tried to stall the press for as long as possible. Finally, Udall, believing that all seven proclamations would be signed on the Friday or Saturday before the inauguration, instructed Boatner to release the information on all seven for Monday's news. By Saturday morning, Jan. 18, the *New York Times*, for one, was aware of the possible proclamations and the information was out over the news wires.

Udall was going to warn Johnson of these press releases at their scheduled Saturday meeting to discuss the monument proposals, but the meeting was postponed. When Johnson saw the releases coming over the news ticker on Saturday afternoon, he called Udall. The secretary recalled, "The president was very unhappy and bawled me out good that he hadn't made a decision and we turned it loose." This was the last official conversation Udall and Johnson ever had.

On that same afternoon, Pierson was also pressuring Udall to help him on a completely separate issue dealing with Venezuelan oil rights. When Pierson implied that the proclamations would have a better chance if the secretary cooperated on the oil issue, Udall was infuriated. He had just been reprimanded by the president, and now his special counsel was telling him to play politics with the monument proposals that had become his personal crusade.

"I've made my last arguments on the parks," Udall told Pierson. "You can do what you damned please... I'm through... I've made my case and if you don't want to do anything on the parks, that's fine."

Udall believed, at that point, he had quit only two days before Nixon's inauguration. He retracted the news releases and went hiking along the Chesapeake and Ohio Canal all day Sunday, his last hours in office. Nevertheless, he had National Park Service Director Hartzog wait in Udall's office throughout the day just in case Johnson changed his mind and decided to sign the proclamations.

Meanwhile, back in Utah, Superintendent Heyder was completely in the dark about what was happening in Washington. After submitting his maps in early October, he had fielded a few clarification questions, yet he had no idea what the final boundary decision was. He didn't even know that the

Capitol Reef National Monument expansion was part of a seven million-acre proposal. If any correspondence or memos were circulating, they were doing so well over the head of the superintendent of Capitol Reef National Monument.

Then, just a few days before the end of Johnson's term, Heyder received an "eyes only" package from Washington that contained the press releases and maps for each of the seven proposed monuments. According to Heyder, there were two different maps of an expanded Capitol Reef National Monument, one with the eventual proclamation boundaries and one which also included the Circle Cliffs and Escalante canyons. There were also two different press releases for the monument's expansion, apparently depending on which one Johnson chose to sign. Heyder did not know which alternative would be included in the official proclamation.

Then, on Friday, January 17, Heyder's wife was driving to Richfield when she heard over the radio that the monument was going to be enlarged. She called the superintendent, who in turn phoned Regional Director Kowski in Santa Fe. Kowski was just as surprised as Heyder. The regional director called back later that afternoon and told Heyder that nothing had been signed. According to Kowski, the news leak had apparently come from the Utah delegation, which had been briefed on the Arches and Capitol Reef proposals the day before. There was still no word about which Capitol Reef National Monument expansion boundary plan had been chosen or what was actually happening in Washington. According to Heyder, at that stage, "nobody knew anything."

On Monday morning, January 20, Lyndon Johnson's last day in office, the president called his special counsel into his bedroom as he dressed for the inaugural ceremonies. Pierson later recalled that the two of them spent an hour discussing the various proposals,

> going over these cases one last time while he was deciding whether or not he would sign any or all of them. [Johnson] finally decided that he would sign the smaller ones and not sign the larger ones.

The president's decisions were made public that morning in a White House news release. The "smaller" monument proclamations that Johnson approved and signed established Marble Canyon National Monument (which was later added to Grand Canyon National Park), and enlarged Katmai in Alaska and Arches and Capitol Reef in Utah. Of the approximately 300,000 acres added to the national park system, the largest portion was attached to Capitol Reef. Even though the seven million acres originally proposed by Udall had been slashed significantly because of size and controversy, the 215,000-acre addition to Capitol Reef would prove extremely significant.

As for Stewart Udall, he believed Johnson had capitulated to unwarranted concerns over the response from Congress....

Udall was, of course, disappointed in Johnson's eventual decision to approve only the smaller monument additions. The last two months of turmoil certainly did not help to ease Udall's disillusionment. Had Johnson signed all the proclamations, National Park Service lands would have been increased by 25 percent. Udall's record as Secretary of the Interior was already very impressive: It would have been remarkable with this last legacy. As it turned out, however, most of Udall's proposals regarding Alaska were eventually included in the 100 million acres protected by the Alaska Lands Act during Jimmy Carter's last days as president in 1980.

Lyndon Johnson's conservation record would have remained impressive even had he refused to sign any of the proclamations. His refusal to sign the larger area proclamations, however, illustrated his political priorities as opposed to the aesthetic values of Stewart Udall. According to historian John Crevelli, Johnson eventually agreed to the 300,000 acres to salvage some of his political prestige and because

> it would be another small step in protecting the natural environment he really loved and in giving the people one last gift in his goal of the Great Society. He needed to go out with love. His ego demanded it.

President Johnson may have given a smaller gift of love than Udall would have liked, but to many native Utahns, the enlargement of Capitol Reef was an outrage. They considered it too big, too surprising, and too much an example of an arbitrary, uncaring federal government. Lyndon Johnson's fears that the proclamations would be controversial were about to be realized. The negative reaction from neighbors and politicians ensnared Capitol Reef in a swarm of controversy that has never been completely resolved.

Yet, if Johnson had not enlarged Capitol Reef National Monument, would the Waterpocket Fold be a national park today? Conflict and controversy are inevitable when land use policy changes, especially when those changes occur so quickly and dramatically. It was now up to Congress, the National Park Service, and concerned residents to wade through the ensuing rhetoric and find legislative solutions to the seeming incompatibility between traditional use and preservation at Capitol Reef.

# 41

## PUTTING TOGETHER THE NATIONAL PARK (1992)

### LURT KNEE (1909–1995)

*Lurt Knee lived at Sleeping Rainbow Ranch, along Pleasant Creek in the park, for more than fifty years. Here is Lurt's take on park expansion. (From interviews by Bradford Frye, 1991, and Stephen Trimble, 1994.)*

I helped put together the national park; it was practically put together here in this house. In this way: when I first came in here I realized that I was adjacent to a national monument, this is what I was *trying* to do. Then my goal was to make it into a national park. I worked with Senator Moss and the Secretary of the Interior, Stewart Udall, and Bates Wilson out of Moab who was the superintendent over three national parks at one time or another. You've probably heard of Bates.

He came over when I first came here, and I said, "Bates, how long does it take to make a national park out of this national monument?"

He said, "Lurt, it will take at least twenty years."

I said, "Well, that's fine."

We went down through here and made the whole survey, we took in certain cliffs and the whole works, and up over Boulder Mountain. When we got up on Chokecherry Point, Bates said, "Lurt, if there is anyone that ever questions that this should not be a national park or wonders if it should be, you take them up here."

The local people didn't really want a park; they fought it pretty hard. But the people up county and up in Salt Lake, they were the ones that wanted it

Fruita history on the walls of the Gifford home, 2015.

into a park, and the people, the county, the business people of Loa, Bicknell, and Torrey wanted the national park.

Your cattleman didn't want it because he was afraid they were going to step on his toes. If you look at the national park today it's all notched out where they met around different interests; they were very fair with them. The ranchers still have the same that they always had, but they were scared that they would lose their cow rights—even if they had one cow on the monument or on the park.

How would you like to try to starve a dollar out of a cow in this country? This is not cow country. But they did it somehow. Well, the reason is that they loved the country and they were afraid that they wouldn't be able to do what they loved, to break loose from up here in the county and take their cattle and go down through here and camp out with them. Wherever they pleased. This is the thing that was hurting them, that they wouldn't be able to do that. They had a love for the country, you see, as well as the cow.

# 42

# RIGHT-OF-WAY (1999)

JARED FARMER

*Jared Farmer included this piece about the Burr Trail in* Glen Canyon Dammed: Inventing Lake Powell and the Canyon Country *(1999).*

The Burr Trail was seemingly made to be a scenic byway. After leaving Boulder, it winds around giant beehives of Navajo sandstone before dropping into narrow, red-walled Long Canyon, which eventually opens up into a huge basin surrounded by the Circle Cliffs. After crossing the basin, the road comes to the abrupt, awesome edge of the Waterpocket Fold, an uplifted monocline within Capitol Reef National Park. In a series of spectacular switchbacks, the road drops from the fold to the blue-shale desert near the base of the Henry Mountains and turns southward toward Lake Powell.

The route takes its name from John Atlantic Burr, a pioneer stockman. Starting in the 1880s, ranchers used the Burr Trail to move their livestock from summer pasture on the Aquarius Plateau to winter pasture in the Circle Cliffs basin, what they called the "Lower Country." In 1882, a local woman commented on the trail: "It is mostly uphill and sandy knee and then sheets of solid rock for the poor animals to pull over and slide down. I never saw the poor horses pull and paw as they done today." In the early twentieth century, a few random prospectors, geologists, archaeologists, and tourists traveled the Burr Trail by mule or horseback. A third option—wagons—opened up in the mid-1930s, following trail improvements. Cars (what few of them existed in Garfield County) couldn't handle it until 1947, when the Boulder

The Burr Trail crosses the Waterpocket Fold. Look beyond the switchbacks to follow the road across the Circle Cliffs in Grand Staircase–Escalante National Monument, 1994.

Cattleman's Association and the BLM sponsored the first bulldozer work. At last, the Burr Trail was a "road." Additional improvements, including the switchbacks down the Waterpocket Fold, date to the uranium boom. "Even before the road was completed," wrote one local, "prospectors in jeeps spread over the country staking claims." Finally, in response to Lake Powell, the county upgraded the road once more and continuously maintained it thereafter.

As the only direct link between western and eastern Garfield County, the Burr Trail became vitally important in the late 1970s, when a mining firm built a new town near Bullfrog. Ticaboo, as it was called, owed its existence to a resurgent uranium market. Its mine and mill accounted for nearly half the tax base of Garfield County in the early 1980s—less an indication of uranium's value than of the county's poverty. Garfield typically had unemployment figures double or triple the state average.

By improving the Burr Trail, the county hoped to revive its economy. A paved road would, it believed, ease cross-county transportation, facilitate mining and mineral exploration, and enhance tourist revenue. The existing road received ten to fifteen traffic counts per day; the county hoped to boost the average to 250. "The paved road will open up this country to tourism," said Del LeFevre, commissioner of Garfield County. The title of the road engineering study said it all: *Boulder-Bullfrog Scenic Road: A Vital Link in the Grand Circle Adventure*. Utah senator Jake Garn championed the project and tried unsuccessfully to secure matching federal funds. The county found a more receptive audience in the state legislature, which approved a preliminary funding bill in 1986, over the objections of environmentalists. Justifying the expenditure, the bill's lead sponsor asserted that federal land regulations "have just about wiped out all potential for economic development [in southern Utah]. We're left with tourism to fill the economic void." Eager to get moving, Garfield County opened the first road bid in February 1987, but only days later the entire project landed in federal court. The lawsuit named four plaintiffs: the Sierra Club, the National Parks and Conservation Association, the Wilderness Society, and the Southern Utah Wilderness Alliance (SUWA).

Barely five years old, SUWA represented a new generation of environmentalists in Utah. Formed in response to the BLM's wilderness inventory process, the group had one priority: preserving the maximum possible acres of roadless land as congressionally designated wilderness. SUWA rejected the philosophy that to preserve wilderness you had to compromise and allow development somewhere else. Though it preferred tourism over other industries, the group believed that the highest use of the land—to be left alone—wasn't an economic use. SUWA had seen enough "improvement" in the canyon country. It fought for a wild Escalante River and an unimproved Burr Trail, not one or the other.

In opposing the road project, SUWA underscored its potential impact on adjacent Wilderness Study Areas and Capitol Reef National Park. But there was a surprising auxiliary concern mentioned in SUWA's literature: the dirt road itself. "There is something beyond the immediate sensory pleasure that makes driving the Burr Trail such a special experience. There is an awareness felt once on the trail that civilization just disappeared in the cloud of dust flying behind you. A clarity of thought emerges, proclaiming that you are truly at one with the earth." This marked a new development in attitudes about roads in southern Utah. In the 1950s and 1960s, wilderness lovers could look past (or even look forward to) new roads because so much primitive country remained. By the 1980s, however, wilderness in the canyon country was perceived as threatened, dwindling. New or improved roads were unacceptable because roads destroyed wilderness. However, certain existing dirt tracks—exceptionally remote or scenic ones—became, by their association with the vanishing wild, objects of nostalgia, even reverence.

Of course, environmentalists were not the first Utahns to associate roads with hallowed ground. Mormons in southern Utah had their own deep feelings about the purpose of roads in wild country. To understand, it helps to return to the Hole in the Rock pioneers and the incredible trail they blazed. In 1940, the Mormon residents of San Juan County organized a commemorative trek to the slickrock dugway at the rim of Glen Canyon. In advertising the event, the local newspaper emphasized the proper attitude:

"While the trip will be most fascinating from the standpoint of scenery and associations, it is not going to be a lark or vacation—but a serious pilgrimage. It will be an earnest and reverent group going to a shrine." This language, though unusually explicit, expresses a cultural viewpoint common in rural Utah. (Garfield County, for example, is over 90 percent Mormon.)

As Wallace Stegner once wrote, "Symbols of the trail rise as naturally out of the Mormon mind as the phrase about making the desert bloom as the rose—and that springs to Mormon lips with the innocent ease of birdsong." In Mormon historical consciousness, the trek across the plains to Utah ranks second only to the restoration of Christ's gospel by Joseph Smith. Besides the prophets, the great heroes of Mormon history are the pioneers, the trailblazers, the road builders. In rural Utah, this reverence for road building has, until recently, been reinforced by personal hardships due to poor transportation. Supplies, medicine, mail, family—all of these were only as close as the roads were passable. Roads were among the most basic and essential forms of progress, and the all-weather highway constituted the ultimate dream, a fulfillment of the good work begun by the pioneers.

To summarize, urban environmentalists looked at the Burr Trail and mainly saw the work of nature surrounding it, while rural Mormons mainly saw the work of people supporting it, improving it. "Since the time Brigham Young sent our ancestors to settle Southern Utah," said Cal Black, former commissioner of San Juan County, "we have been told and have worked and waited for the time when our area, in which most of us here were born, and have remained by choice, would be opened up so that all people could come and enjoy the scenic beauty that we have always known and loved. This has also meant, in our dreams of the future, that we would be more easily able to make a livelihood for our families, to provide them with the educational and other opportunities our City cousins have always taken for granted." Case in point: in the 1980s, Margie Lee Spencer of Escalante drove the Burr Trail twice a week for nine months of the year to reach her job in Ticaboo. She taught school. When questioned as a court witness, Spencer said she met all sorts of drivers on the road: health officers, tourists, ranchers, boaters, prospectors, sheriffs, insurance salesmen, backpackers. They came in compacts, trucks, vans, campers. Not all were prepared; Spencer pulled more than one station wagon out of the mud with her Ford Bronco. "I've passed cars with Alaska license plates," she said. "I stopped and talked with people from I swear almost every state in the union." When asked if she believed the proposed road improvements would "take away from the wilderness characteristics" of the area, Spencer replied, "I don't see how it could. It would just make it so there weren't deep ruts in the road and so that people didn't slide off the road. And it would just make it so people could actually see more of it."

Isn't that what environmentalists wanted, too? After all, three of the four plaintiff groups had earlier testified in support of the Canyon Country Parkway [*a circle route proposed to bolster tourism in the 1970s that would have linked Kanab, Escalante, and Torrey with paved roads, including paving the Burr Trail*]. When confronted with that history, however, environmentalists dismissed it as irrelevant because fifteen years had passed since the hearings, and in the meantime the National Environmental Policy Act and the Federal Land Management and Policy Act had become law and now applied to the area. Before any roads were paved, the wilderness groups wanted to see a comprehensive regional plan for tourism. Garfield County rejected the idea. Understandably, it believed environmentalists were employing a double standard where they had no business in the first place. At one point in the Burr Trail controversy, citizens of Escalante became so frustrated by outside interference they hung the directors of SUWA in effigy.

The conflict came to a temporary resolution in 1987, when a U.S. district judge ruled that the county had a right-of-way to at least the first twenty-eight miles of the Burr Trail, from Boulder to Capitol Reef National Park. An appellate court later affirmed that decision, and road work began in 1988. Soon after, four of the county's bulldozers were sabotaged by one disgruntled environmentalist, accomplishing nothing. Work proceeded to the western boundary of the park, beyond which the right-of-way was still in question. When the Burr Trail reopened, it was partly surfaced, partly dirt. No one was happy.

I remember the Burr Trail. I first read about it when the lawsuit by environmentalists failed. Immediately I planned a driving trip. I had to arrive before the bulldozers, and I made it just in time. Actually, my dad did all the driving, since I was underage. I merely had to stare. It was a perfect October day, warm, vivid, a day for breaking hearts. Once back home, I wrote an essay, the first time I'd ever felt compelled to record something in words. I'd been on a voyage to a new world for the first and final time. Never had I seen so much beauty: the foreknowledge of loss gave me unprecedented, passionate awareness.

Graffiti on bridge over the Gulch, Burr Trail, 1997.

It was four years before I could return. When I went, I drove at a crawl on the asphalt, looking for rocks and feelings I remembered. I couldn't find them. I missed the dust, the clarity. I drove away depressed.

Next time I saw the improved Burr Trail was with friends who'd never been. To them it was beautiful, and looking through their eyes I had to admit it was—still, it wasn't the same. On a brand-new concrete bridge I spied some sympathetic graffiti: "KEEP IT LIKE IT WUZ!"

Of course, you can only keep things like they are, not like they were. That's the poignancy of the spray-painted message: it speaks to the paralysis of nostalgia and the bitterness of hindsight. It didn't occur to me until later that people in Boulder might be nostalgic, too—not for the rough road itself, but for some of the rural qualities safeguarded by roughness.

Sulphur Creek Narrows, 2016.

# PART VIII

## REFUGE

### In the Wild Heart of the Fold

What brings us to national parks? As tourists, checking off a park on our list of must-sees. As travelers, soaking up the ambience of each preserve. As adventurers, looking for the thrill of remote wilderness. Even in the twenty-first century, Capitol Reef and its surrounding country, from Boulder Mountain to the Henrys, offers sufficient solitude, wildness, and challenge—and accessibility—to lure a surprising diversity of visitors.

Writers and artists rely on Capitol Reef for the spark to fuel their creative responses to the landscape. Their goals are emotional, their stories personal. They look inward and outward, balancing the two in singular and revealing voices. And so we walk down the trail with redrock aficionados who come here for many reasons, seeking to understand their relationship—and ours—to the heart of the Colorado Plateau.

Art historian James Swensen gives us a critic's insight into an innovator's mind and heart. Minor White was a major twentieth-century photographer and teacher whose signature image comes from Capitol Reef. Swensen helps us make sense of White's cryptic and abstract detail of a Moenkopi cliff. I also take us on a quick tour of other photographers and artists who have done important work in the park.

Capitol Reef can restore as well as inspire. Ann Whittaker writes a letter to her grandfather, reflecting on his pilgrimages to Cathedral Valley and the gift of Capitol Reef to "swallow sorrow." For several years leading up to the Great Recession of 2008, wilderness therapy programs were the biggest employer in Wayne County. The excerpt from Gary Ferguson's *Shouting at the Sky* captures the power of this land for the young people who came here to heal. Greg Gordon led field studies trips for university students for years, and in this excerpt from *Landscape of Desire*, we meet some of his students, complete with their quirky trail names, and we see how they react to "civilization"—the Capitol Reef campground—after time in the backcountry.

Park rangers occupy a middle ground between science and story. Both Rose Houk and I worked as seasonal rangers before we became writers, and we bring to our adventuring an interpreter's passion for content. Here, each of us chooses to end our books about Capitol Reef with canyon journeys, Houk's trip down Halls Creek in *Canyon Country Eden* and my trip through a universalized side canyon based on that same remote stretch of the park, at the close of my 1978 booklet, *Rock Glow, Sky Shine*.

Today's writer-travelers are canyoneers and backpackers and wilderness guides. Craig Childs, another writer in love with the canyons and deserts, grapples with the effects of technology on his wilderness experience, in this case, a cell phone link to home. Longtime backcountry guide Steve Howe, who lives in Torrey, takes us on a challenging transect over and through and across the Waterpocket Fold. And Charles Riddel trades fancy technology for the simple magic of a bicycle, pedaling through Capitol Reef and allowing that self-powered vehicle to reawaken his sense of wonder.

# 43

## CAPITOL REEF'S MOST FAMOUS PHOTOGRAPH

### Minor White's *Moenkopi Strata, 1962* (2016)

JAMES SWENSEN

*Art historian James Swensen, who teaches at Brigham Young University, has made the history of photography in Utah national parks one of his specialties. This work led to his fascination with one remarkable image taken in 1962 in Capitol Reef by Minor White.*

If there is one photograph emblematic of Capitol Reef National Park, it might be Minor White's image *Moenkopi Strata, Capitol Reef, Utah, 1962*. This may not be the first photograph that comes to mind when thinking of this stunning preserve along Utah's Waterpocket Fold, but it is one of the most well known. Although cryptic, it also provides insight into this different place and the unique experience of its space.

A rich codependency has always existed between photography and the national parks. When "America's best idea" was born in the nineteenth century, photography was in position to bring it to light. By the twentieth and twenty-first centuries, millions, possibly billions, of photographs have been taken of these federally protected sites. Indeed, it is hard to imagine national parks without photography. Sure, paintings could capture their beauty and grandeur. (Think, for example, of the great canvases of Thomas Moran.) Yet there is something distinctive about photography. No other medium can compare with its ability and power to propel people to go, to see, and to share—an essential element of the national park idea.

Photography, in turn, has benefited from the national parks. For more than a century, key landscape photographers have become linked to these places. This is certainly true of Ansel Adams's and Carleton Watkins's connection to Yosemite, and William Henry Jackson with Yellowstone. The vast majority of photographic images in the parks, however, have been made by thousands of amateurs and snapshooters who have spent millions of dollars on cameras, film, tripods, gigabytes, and those damned selfie sticks in an attempt to capture their experiences in the parks. (How many have you made?) At this very moment someone with a camera is peering over the edge of the Grand Canyon, photographing distant Half Dome from the Wawona Tunnel parking lot, or waiting for the right light to hit Delicate Arch. The mere presence of these images residing in photo albums, dusty slide carousels, or as bytes in the ether of a hard drive help visitors hold on to some of the magic of their time in "nature." These preserved photographs act like a visual Leyden jar, providing a spark to the monotony of everyday life.

The area that is now Capitol Reef National Park was first photographed in 1853 when Solomon N. Carvalho hauled his cumbersome daguerreotype equipment across the West as part of John C. Frémont's ill-fated fifth expedition. Since that time the park has continued to attract legions of camera-toting visitors and talented artists and photographers like Ansel Adams, Philip Hyde, John Pfahl, and Stephen Trimble. With all the images that have been made, however, Capitol Reef does not have a photographic icon or even a defined iconography. This is not necessarily surprising.

Capitol Reef is photogenic, but one must work to uncover the breadth of its beauty. Within its boundaries one finds stone arches and bridges, spectacular rock towers, red rock cliffs, and stunning views, but many of these are hard to reach. With the possible exception of the Temples of the Sun and Moon, none of its features are as well known or celebrated as those in the other western national parks. The beauty and spectacle of Capitol Reef resides in its subtlety and the intimacy of its rock forms and tight meandering canyons.

One artist remembered for photographing Capitol Reef is the intriguing Minor White. While not as well known as many of his friends like Adams or Dorothea Lange, White is one of the most important and influential photographers of his generation. Born in Minneapolis in 1908, he purchased his first camera after graduating from college and quickly dedicated his life to the growing medium. Throughout his long career, White was known for creating enigmatic imagery and for his immaculately crafted prints. He was also profoundly influential as a teacher and in his publishing endeavors. His most important contribution, however, may be his tireless efforts to find meaning and depth in the photographic image. White's belief that photographs have soul drove every aspect of his multivalent pursuits.

White came by his quest naturally. He was a believer, a mystic, an intellectual, a searcher, a guru. Always in pursuit of higher meaning, he converted to Catholicism during World War II and later became an ardent supporter of Zen Buddhism, astrology, and the teachings of the Russian philosopher G.I. Gurdjieff. He employed a host of strategies to unlock inner truths within a photographic image, including semiotics, symbolism, and even hypnosis. For White, the object that appeared in a photograph was never straightforward. It was a reflection, an "equivalence," that provided a window into the inner workings of the individual. For White it was not a question of what you see, but "Where does it strike you?"

As part of his pursuits Minor White searched out locations that offered an opportunity to explore a metaphorical world. In this he saw himself as "The Wanderer" constantly on a quest for essential knowledge. Encouraged to visit what was then Capitol Reef National Monument by his friend, the photographer and surrealist Frederick Sommer, White traveled to the future park in 1961 and returned each summer for the next five years.

This landscape of stone and light clearly resonated with the photographer. White, then in his fifties, often traveled with younger companions whom he nurtured and who provided youth and energy to their joint pursuits. On his first visit to Capitol Reef he traveled with Bill LeRue, a young schoolteacher and workshop assistant. One year later White spent a week camping out of his van with Drid Williams, a modern dancer from New York City. Williams remembers taking "hundreds of photographs" while exploring the area. She was not a photographer any more than White was a dancer, but during their time together they learned from each other and were united in the "similar quest of nurturing and developing the spirit."

Photographs from White's first two trips to Capitol Reef became part of "Sequence 17," a series of disparate images organized in such a way as to

Minor White, Utah. Photograph by Arthur Lazar, 1965. Used with permission of the photographer.

create deeper meaning within the viewer. White insisted, "While rocks were photographed, the subject of the sequence is not rocks, while symbols seem to appear, they are pointers to the significance." He continued, "The meaning appears in the space between the images, in the mood they raise in the beholder. The flow of the sequence eddies in the river of his associations as

Minor White's (1908–1976) *Moenkopi Strata, Capitol Reef, Utah, 1962*—an image James Swensen calls "Capitol Reef's most famous photograph." Used with permission of the Minor White Archive, Princeton University Art Museum.

he passes from picture to picture. The rocks and the photographs are only objects upon which significance is spread like sheets on the ground to dry."

When White returned to Capitol Reef in 1965 he brought along Arthur Lazar, a young friend and photographer who also knew the area well. Lazar was dismayed to find that White was more interested in "going up to the

Mountain" to meditate than taking pictures but eventually convinced the photographer to pick up his camera and they spent the better part of two weeks in the sparsely populated national monument. During their time together they worked early in the morning and in the late afternoon to capture the right light. Lazar showed White his favorite spots and the older artist returned the favor. Together they explored well-trodden sites like Grand Wash, Sulphur Creek, and Cohab Canyon, as well as many spots off the beaten path. In one hard-to-reach ravine, which White called "Mass Canyon," the team photographed swirling forms and abstract gestures in the glossy, jet colored mud at their feet. Despite White's eccentricities and rituals, Lazar remembered his experiences in 1965 as deeply influential and "wonderful."

It was during White's second trip with Drid Williams that he created *Moenkopi Strata, Capitol Reef, Utah, 1962*. White's title is important, revealing the what, where, and when of the photograph. In this it is strangely descriptive in a geological sort of way. The Moenkopi formation, a layer of sedimentary stone, was formed 240 million years ago during the Lower Triassic period. Widely exposed on the Colorado Plateau, the Moenkopi is characterized by evidence of flowing waters like ripple marks, channel deposits, or mud cracks. Looking at White's image, however, it is not clear exactly what one is seeing. Like certain elements within modern dance, it juxtaposes the lyrical and the jagged. Deep, darkened recesses balance slashing, curving forms of white and grey. With no sense of scale, the rock seems both distant and near. It is bewildering, disorienting, and menacingly anthropomorphic. Yet the question remains: Is it rock, or water, or mud, or ice? There is no way of knowing, and *that is exactly the point*.

In his work, Minor White was careful in referencing the physical world. He wanted the viewer to strain, ponder, and question what is seen and expressed within a photograph. According to Gurdjieff's teachings, "most of us are asleep most of the time, experiencing the world through a disastrously fragmented and deadened intelligence, and that the first order of business is to wake up and stay awake." In this way White's images are never passive. Exploring this photograph was meant to be jarring, shaking the viewer into consciousness.

Despite the esoteric challenge inherent in *Moenkopi Strata*, the "things" White captured on film exist in some way in the physical world. No matter how cryptic the photograph is, it is *of* Capitol Reef. With images like this White believed that "the spring-tight line between reality and photography has been stretched relentlessly, but it has not been broken." This is certainly true of this picture. Although abstract, this image is an example of what White called an "extraction," or an "isolation" from "the world of appearances."

While difficult, maybe even impossible, to perceive exactly what White photographed in 1962, he did capture essential elements of Capitol Reef. Although he worked in black and white, he shows a range of textures. His "Moenkopi strata" is opaque and transparent, striated and smooth, sharp, rippled, and rough.

Like White's enigmatic photograph, experiencing the park can be surreal. In the right light, haunting, sometimes humanlike, forms often surround the visitor in rock hoodoos and decaying sandstone. Hiking through Capitol Reef's canyons and along its rocky plateaus can be disorienting. There is a push and pull between distant views and the intimate forms at one's hands and feet. It does not take long to realize that the subtle variations in stone and vegetation, however slight, can be just as sublime as the park's most spectacular vistas. In this way White's image might not be as strange or as abstract as one might originally think. *Moenkopi Strata, Capitol Reef* might reveal *how* one experiences the park more than specifically describing *what* is found within the park.

Since its creation more than a half century ago, *Moenkopi Strata* has become one of White's most famous images. It appears in art history textbooks and in the histories of photography. The print is also found in important museums and photography collections around the world. Google the title to follow the perambulations of this photograph. No matter where it goes, physically or as pixels, White's image needs to be connected to its place. Both deserve attention. Drid Williams believed that Minor White had an ability to "create spaces where truly significant things could occur." This is true of the time one might spend in this photograph and particularly in that place called Capitol Reef.

# 44

## CAPITOL REEF, ILLUSTRATED

STEPHEN TRIMBLE

This landscape of rock yields endless abstracts and details. Like Minor White, Bruce Barnbaum is a major photographer whose Capitol Reef images startle and intrigue. He is a master printer, and his black-and-white prints from the park rival White's for provocation and strength.

More experimental photographers have made images just as indelible. In a series of 1977 photographs, John Pfahl added human objects to landscapes to make a philosophical statement about belonging. "What Pfahl obtains is nothing short of the image and the idea at the same time: both rigorously intellectual and lushly sublime," critic Anthony Bannon wrote of Pfahl's *Altered Landscapes* in a 1979 essay. Two of the important images in Pfahl's sequence come from Capitol Reef. *Moonrise over Pie Pan* is my favorite.

Philip Hyde and Eliot Porter were the great conservation photographers for the Sierra Club "battle books" of the 1960s and 1970s—using photos to advance conservation. Both worked in Capitol Reef, mostly in color. Hyde made images in the park for his classic collaborative book with Ed Abbey, *Slickrock*. The image of strikingly striped rock in a side canyon reproduced here (on page 287) opens the Waterpocket Fold portfolio in *Slickrock*.

I have often quoted Philip Hyde's preface to *Slickrock*, in which he articulated the wilderness photographer's fear:

> The focus of this book is on a part of Earth that is still almost as it was before man began to tinker with the land.... Telling thousands about it—to get their help in what must be a prolonged struggle to

Bruce Barnbaum: *Boulder and Metamorphosis Wave, Capitol Reef.* Used with permission of the artist.

> keep it wild—is a calculated risk.... I have some hesitation in showing more people its delightful beauty—hesitation born of the fear that this place, like so many others of great beauty in our country, might be loved to death, even before being developed to death. So, if our book moves you to visit the place yourself sometime, first make sure you add your voice to those seeking its protection.

For every place, Philip Hyde said, "there will always be people that want to exploit it, and there will always be people—hopefully—that want to save it and keep it as it is." Better to publish your photographs and rally the troops. What's in the frame of the photograph matters artistically, to be sure, but what's outside the frame can destroy it.

The generation that followed—including myself—knew that Philip Hyde's photographs helped save Grand Canyon from dams on the Colorado River. And that Hyde's photographs of the Escalante wilderness helped

John Pfahl: *Moonrise over Pie Pan, Capitol Reef National Park, Utah, 1977*. Used with permission of the J. Paul Getty Museum, Los Angeles.

save the Kaiparowits Plateau from destruction by coal mines when President Bill Clinton proclaimed Grand Staircase–Escalante National Monument in 1996. We know the power of nature photography. And we have tried to live up to this legacy.

Eliot Porter, Hyde's contemporary, did equally important work as a conservationist and photographer. Porter came to Capitol Reef when I was working in the park as a ranger. One hot afternoon humming with gnats, vibrating with silence and sun, the consummate large-format landscape photographer and his assistant walked through the Visitor Center door, grateful for the shade and freshly astonished by the rough roads in Cathedral Valley. Porter asked the ranger—me—for ideas for photo locations. He was working on a book to mark the 1976 US bicentennial, a book celebrating the American landscape from Maine to the Pacific coast, with Porter's photographs and text by Wallace and Page Stegner, father and son. With his patrician roots, wire-rimmed glasses, and field khakis, Porter came across

Philip Hyde (1921–2006): *Canyon in Waterpocket Fold, Capitol Reef National Park, 1970*. Used with permission of the Estate of Philip Hyde.

as organized, intense, and intellectual. He had trained as a bacteriologist at Harvard before dedicating himself to photography.

I was beside myself. Porter and Stegner were my heroes. But of course I tried to be mature and matter-of-fact. Professional. Porter, to his credit, treated me with more respect than I deserved—deferring to my local knowledge though by then I had booked only a couple of months in the park. During the next week, I encountered him again in the backcountry and continued to steer him toward my favorite places.

I watched for the book for years. I discovered it newly published in 1981. Plate 74 in *American Places* is a picture of cottonwoods at the humble waterhole of Fountain Tanks—a classic Porter composition framed by the angles and curves of the golden sandstone—that I assumed Eliot Porter added to his portfolio because one searing June day I sent him and his assistant hiking in that direction from The Post. (I was disappointed when I learned the image in the book came from Porter's 1963 visit to Capitol Reef.)

Eliot Porter (1901–1990): *Pool and Willows, Waterpocket Fold, Utah, August 21, 1963*. 

Guy Tal: *Break in the Storm*. Used with permission of the artist.

In the 1960s and 1970s, The Post was still a pretty remote place—a road junction marked by a simple sign along a dirt road in the southern reaches of the Waterpocket Fold canyons, thirty-five miles from pavement, more than one hundred miles from an interstate. Where else was I going to run into Eliot Porter but in a place like this—just the sort of place he sought for his pictures? The encounter seemed perfectly neighborly.

Today, photographers stream through Capitol Reef country. Some choose to stay, including Scott T. Smith, who uses his art to campaign for the protection of dark night skies, and Guy Tal. Born and raised in Israel, Tal lives in Torrey, and he's made Capitol Reef country his home landscape in photographs filled with an appreciation for magical and subtle light.

Tal writes of his choice of Capitol Reef country as his photographic subject in his book *More Than a Rock*: "I wanted more than just beauty and more than just images: I wanted the comfort of open desolate places; I wanted naked rocks and shrubs and landmarks whose names I knew, and whose stories—at least some of them—were familiar to me.... I realized how much my own relationship with my subjects, their stories woven with mine,

V. Douglas Snow (1927–2009): *Capitol Reef*. Scott M. Matheson Courthouse, Salt Lake City.

and in a lesser way my relationship with the medium of photography, have made my own art more significant by virtue of making my life experiences more significant. Try as I might, I cannot treat my experiences and my art as distinct and separate things."

Such creativity and emotion, both tender and fierce, fueled Doug Snow's landscape paintings. One of Utah's preeminent artists, V. Douglas Snow built a home in Wayne County in the 1960s, immersing himself in his subject. He lived below the Cockscomb at the base of Boulder Mountain in Fish Creek Cove, a glorious outcrop of Navajo Sandstone nearly included in the original Capitol Reef National Monument in the 1930s. He painted this stegosaurian ridgeline over and over, as Cézanne painted Mont Sainte-Victoire, through the fullness of his life. Doug Snow transformed Capitol Reef's slickrock skylines into art. In a lecture about his work, he revealed a key to understanding his mural-sized paintings: think of the swirl of colors at the top as the sky and the grounding forms at the bottom as the land, Snow said, and imagine everything in between as emotion.

Another window into Snow's understanding of Capitol Reef comes from his unpublished play "Blind Sight." His lead character, a painter who speaks for Snow, says, "I've really painted what I see, and what I see has been increasingly influenced by what I feel about what I see."

Snow painted dozens of landscapes that captured this confluence of place and emotion. To his chagrin, one became controversial—a major work that rises behind the bench of the Utah Supreme Court in the Scott M. Matheson Courthouse in Salt Lake City. The Utah Arts Council commissioned Doug Snow in 1997 to create *Capitol Reef*, a mural nineteen by fifteen feet, to illustrate conflict and resolution. Snow wrote, "An apocalyptic thunderstorm saturating the slick rock sandstone, followed by intense heat from the summer sun, brings about a magical moment. Vaporous clouds rise from the cliffs and domes of southern Utah's desert country, providing a powerful and poetic experience of hope and renewal."

Once Snow's work was installed behind the bench, the justices concluded that the huge painting had too much "apocalypse" for their tastes. They thought the painting would be distracting to lawyers. A gray curtain now covers *Capitol Reef* when court is in session. That sexy landscape is just too striking for those staid urban jurists.

# 45

## SHOUTING AT THE SKY (1999)

GARY FERGUSON

*Montana writer Gary Ferguson has written about rivers, wolves, grizzlies, and wild places in more than twenty books. This excerpt from* Shouting at the Sky *comes from his time with students at the Aspen Achievement Academy, a wilderness therapy program in Capitol Reef country. Years later, he described his fieldwork for the book as "the most significant personal experience I've ever had."*

The middle of Utah's Red Desert, two, maybe three in the morning. Light from a full moon is spilling all over the place—down the shoulders of Caineville Reef, across the long, flat sweeps of sage and rabbit brush and greasewood, through a thin braid of dry, nameless washes, onto the faces of seven teenage girls scattered across the ground at the edge of a box canyon, hoping for sleep. Lisa and Jenna are having weird dreams again, twitching and mumbling, setting off on what seem to be conversations, passing off a slur of words and grunts, even instructions: "Not that way," Jenna is saying. "Go left. It's over there." I'm trying to remember it all, give the words back to them in the morning on the off chance they might hold some kind of meaning.

A few feet away Nancy is poking Tricia in the ribs, trying to get her to stop snoring. And beyond that, under the only tree within a half mile, is the new girl, Brenda, the one who just went on suicide watch. Such a strange routine for her now: the drawstring taken from the hood of her sleeping bag, the laces pulled from her boots—just in case. A blue tarp spread tight

Red Desert by moonlight, 2018.

across her bag with two staff lying on the outside edges, one on either side of her, the better to feel every move she makes.

If this were your first week in the field, you'd probably be thinking it'd been one hell of a day. Carla had her first try at leading the group on a night hike, laying open a personality that was something like a cross between Joan Crawford and an aerobics instructor. No one took it well. And then that thing with Brenda. Around midnight she decides to sit in the middle of the road, refuses to walk any more, starts cursing, spitting on us. Says what would really make her happy is if all of us would burn in hell. Before it's over she manages to raise everyone's hackles, even stalwart Jenna and the normally tranquil Nancy, sending them over to staff with teeth and fists clenched, asking if they could "hit her just once. C'mon, just once!" Not surprisingly, when we got to camp, sometime around one in the morning, Lisa called a group to talk about it. But instead of telling Brenda what a bitch she was—and that would have been pure Lisa—she said only that she understood Brenda's loneliness, that she knew about the anger she felt at being here. That things would get better. That if she needed to talk....

So much craziness, and on one of the most beautiful nights I've ever seen. The kind of desert night that feels like a gift. Delicate. The air filled with the smell of juniper and sage, desert holly and cliff rose, sandstone and alkali dust. And out beyond camp hundreds of sego lilies, one to a stem, their ivory blooms glowing in the wash of the moon. Lying awake through these wee hours I'm thinking maybe all this wouldn't be so hard to get my head around—wouldn't seem so filled with contradiction—if my culture hadn't spent the last hundred years thinking of the wilderness mostly as some kind of tonic: sedative, blood-pressure medicine, speed. The wilds as the place we go to smell the pine and the rain, dangle by ropes from the chins of mountains. It's been such a long time since deserts and woodlands were places of confrontation, stages on which to wrestle with shadows and cry for visions, holy lands hiding strengths that go unsuspected in more common hours.

The old people of this place, the Paiute, knew full well that beauty and craziness would be found together like this, standing hand in hand. Paiute creation myth tells how long ago the earth was danced by two brothers, Coyote and Wolf. Wolf with his perfect, wholesome vision of the world, a creator who never wanted anything more than an abundant life for the people, a life free of anguish, free even of death. And the younger Coyote, spoiled, mischievous, the glib talker who time and again pulled his older brother away from those plans for perfection. And how after a time Wolf went away, leaving the world to unfold according to the imaginations of Coyote. We cast our fate with Coyote, said the Paiute. And so our lives are driven by this strange mix of urge and shadow, by schemes going out into the world meaning to be clever, coming back full of pain.

Now these smudged, sweat-stained girls, kids who never knew of or cared a damn about the Paiute's brother gods, lying here in the shadows of these same ancient canyons, wrestling with Coyote things of their own. With their habits of crack and speed and crystal meth. With late-night trips to the police stations, to the streets, to the suicide wards. Early in the afternoon with girlfriends at school, in the toilets, throwing up lunch. Hammering together pieces of whatever's in reach, trying to survive. Like Nancy a couple of days ago, walking down that dusty trail, talking about her bulimia: "How could I deal with things if I didn't throw up?" she said. "What else is there in my life I can control?" Then later, around the fire, before bed, she starts rapping on the bottom of one of the tin cans we use to cook in, and then someone else starts in with her thumbs against the bottom of her blue metal cup, and

Cloud over the Waterpocket Fold, 2018.

then three more cups and a pair of wooden spoons, until there's this heady thrum drifting out across the desert—in some moments disjointed, but in others, perfect. And right in the middle of it a coyote comes up to the edge of the bench that runs along our camp to the south, gives three bright barks, turns, and walks away. All of us sitting there looking at one another, amazed; never slowing the rhythm, though, never stopping that drumming.

"That coyote," Nancy says over breakfast the next morning, as if only then was it proper to speak of it. "It was awesome." And she slipped that memory into her pocket, and she's been walking with it ever since, all across this empty desert, drinking from it like a spring, smiling over it when the weight of her pack and the long black nights start pressing on her shoulders....

Late afternoon brings clouds—cold and gray, like they were spun out of steel; while so far the most they've managed here in the desert is a few drops of rain, from the looks of it they're losing no end of fresh snow on Thousand Lake Mountain. Just when we're sure the storm is going to get us after all, two big fingers of blue sky crack open in the west to light a band of mist lying along the massive domes of Capitol Reef, turning them into sun-drenched turrets and ramparts, one minute half hidden and the next rising out of the gray like portals in the mists of Avalon. It's simply incredible, a show of beauty that completely overwhelms us, and before long Megan and Jonathan are running back and forth across the hills around camp, gaining this rise and then that one, trying to see it from every possible angle. In the last of the light a rainbow begins to form barely a half mile away, one end planted at the base of Rabbit Butte and the other just to the east, in the foothills of the Henrys. It comes in stages, a few weak bands at first, swelling to something fat, complete, every color clear, as if it had been finger painted on the sky. Even Susan, who by now has shuffled back to her bag, once more sad and dejected, comes hurtling out from under her tarp. "Wow! That's the biggest rainbow I've ever seen in my life!" she says, grinning for the first time in two days....

There's no trace of the six inches of fresh snow that fell on the Henry Mountains five days ago; only the steep, west-facing slopes hold any sign that winter was here at all. As we spin down into the desert, patches of locoweed start to appear, their clustered, ivory blooms scattered across the dry lands like puddles of spilled milk. Desert holly is full on now, as are the first of the evening primroses—exquisite, chalice-shaped blooms hanging from middling-looking vines in places with no names.

Spring in the Utah desert is mostly a roller coaster, full of surprises, sort of like a grizzly bear about to clamp his jaws around your head and then deciding instead to cuddle you and lick your face. You sit there trying to enjoy the wonder of it, but always in the back of your head knowing full well that the slightest thing could set him off again. Even the intrepid Mormons, with their uncanny knack for squeezing life out of inhospitable places, have been at various times brought to their knees in this part of Utah, rocked from one extreme to another—searing drought in one season, flooding in the next. After devastating spring floods in the latter 1800s and again in 1909, the church's leader for the area called the faithful to Loa and in an almost unheard-of move, gave them honorable release from their lives here, praying to a generous Father that he bless the people wherever they go, whatever they do. "Someone has said that one man's heaven is another's horror," wrote a local historian. "So it is with Wayne County. A heaven on earth to some, to others, discouragement, disillusionment, despair, disaster, desolation, and death."

Several field staff have suggested that something very different happens to kids when they're in this desert, compared to when they're high in the forests of Boulder Mountain. More gets done in the desert, they say. The changes seem to be faster, deeper. Most think it has something to do with the fact that there are so few places to hide, no leaf canopies or thickets or tree trunk to offer refuge. Land as a mirror held up to a kid's life—at first, harsh as his worst fears, but over time, at least for some, bright and fierce and wild as youth itself. A land of pain and clarity.

# 46

## SO WHAT IF WE ALL WANT REDEMPTION (2016)

ANN WHITTAKER

*Writer Ann Whittaker lives in Salt Lake City, where she has worked as the content manager of the online travel site utah.com and the designer-curator of redrockstories.org. She "navigates landscapes between corporate, environmental, social, educational, and cultural ecologies." Her writing is fresh, self-disclosive, and compassionate—like this piece about the healing power of Cathedral Valley.*

Dear Grandpa,
Can the desert rescue you from night terrors, make you stop throwing punches in your sleep? Can a temple of the sun, a lifeless sandstone mass, erase what you saw in Germany—burnt down churches filled with bodies? Can millions of years of beautiful erosion in the earth resurrect your wife's brother?

You tell me about a desert landscape where sandstone castles and chimneys tower above the road. You found refuge there, sleeping many nights under the moon and the Milky Way. You drove hundreds of miles out of your way to find rest from dreams of bombs and death. You talk about the desert. You only dream about the bombs. You keep returning to the Temple of the Moon, praying for redemption.

Can a grand wash, a river bed, cleanse a haunted soul? Can a cottonwood tree teach you how to weep like the black desert varnish on canyon walls? Can the song of a canyon wren teach you to speak softly to your children?

Temple of the Sun, Cathedral Valley, 2016.

You tell me, your granddaughter, that once you stopped to help a local farmer mend his fence on your way through a tiny town called Grover. You told me it had rained for days in that desert place, and the mud was thick. You helped a farmer because you grew up a farmer in the west before you became a soldier avoiding barbed wire meant for your decapitation.

Can the sound of water in the desert fill a broken heart? Can a rip in the earth's surface, a grand monocline, swallow a sorrow deeper than the earth's core? Can the moon rise high enough over Entrada monoliths to light our way?

You tell me it can.

I believe you.

Love,

Your Ann Marie

# 47

## LANDSCAPE OF DESIRE (2003)

GREG GORDON

*Greg Gordon teaches environmental studies at Gonzaga University. For twenty years he taught field studies courses for the Sierra Institute and Wildlands Studies, often in southern Utah. This excerpt comes from his book about these field experiences in and around Capitol Reef,* Landscape of Desire: Identity and Nature in Utah's Canyon Country.

*In his preface to the book, Gordon writes, "I've witnessed how extended time spent living in a small group in the wilderness engenders a fundamental shift in the way we regard ourselves and our place in nature… more than anything else the wilderness teaches us how to live in community, as citizens of the biotic community as well as a human community that we actively create."*

While inclusion in the National Park Service spelled the end of the settlement of Fruita, it ironically is now the best preserved example of Mormon frontier life.

Survival in this harsh land depended upon religious devotion to production. The name Deseret is the name for honeybee in the Book of Mormon, and the beehive is the state symbol and industry is the state motto. "In Mormon doctrine, earthly labors carried a direct connection to spiritual progress; one's exertions in the material world directly reflected one's spiritual standing," wrote historian Patricia Limerick. Thus, wilderness can be viewed as antithetical to Mormon philosophy, preventing industrious work such as coal and uranium mining, oil and gas extraction, and power plant production. It threatens livestock interests and can't be exploited for tourism.

Leaving behind the pastoral lands of the east, Mormons were confronted with a grey desert that "did not become beautiful until the second or third generation when it began to be seen through the eyes of those with full stomachs, reliable water supplies, comfortable homes, passable roads, and hearts buoyed up by hope," explained Lyman Hafen, a fifth-generation southern Utah Mormon.

Possibly the isolated landscape informed cultural attitudes, just as the land itself was being affected by cultural values. This history, combined with a sense of ownership of the land, makes many Mormon communities suspicious, if not outright hostile to the idea of wilderness designation. While the church no longer appoints political leaders, it still exercises powerful influence over Utah politics. All of Utah's congressional delegation is Mormon, as is more than ninety percent of the state legislature.

Although the Prophet Joseph Smith "taught the sanctity and unity of all living beings, and that plants and animals had souls," the LDS Church is formally committed to inaction in terms of environmental positions. Yet, many urban Mormons actively support wilderness, including the new director of Southern Utah Wilderness Alliance, a great-great grandson of Brigham Young. And in 1976 church president Spencer Kimball stated, "But when I review the performance of this people in comparison with what is expected, I am appalled and frightened. Iniquity seems to abound.... I have the feeling that the good earth can hardly bear our presence upon it.... The Brethren constantly cry out against that which is intolerable in the sight of the Lord: against pollution of mind, body, and our surroundings."

We spill out of the van onto the lush grass of the campground. Everyone stands around in a daze as if they've just landed in some exotic country.

"Uh, where's the water?" Mud asks.

"Over by the bathrooms, there's a spigot."

"Bathrooms?" says Seeker somewhat mortified. "You mean we have to use bathrooms?"

I shrug. "Suit yourself."

"Can we drink the water?" Mud asks.

"I'm sure it's liberally dosed with chlorine."

Everyone returns from the bathrooms excited and full of stories. "Dude, there was some guy in there washing his underarms," says Yucca.

"These girls had blow-dryers and were doing their hair and putting on makeup," says Sage as if this were truly bizarre behavior.

Huckleberry sits on a picnic table staring past the RVs, green lawns, and twenty-four hour sprinklers to the fujichrome backdrop of azure sky, towering orange cliffs, and purple Chinle slopes.

"It looks fake, like a postcard, or it's behind glass or something," he mutters.

"It's almost too beautiful, but it feels like I can't get out and explore it," Patience concurs.

"That's because it's scenery, not nature," Seeker adds.

"They have big metal signs telling you what to expect on your hike, how far it is, what the trail's like, if it's easy, moderate, or strenuous. Sort of takes all the fun out of it," says Huckleberry.

"There's a Coke machine over there. You can get your Coke and go and check out the wild animals," Bobofet indicates the deer ambling through the

Mule deer in Capitol Reef campground orchards, 1992.

campground munching contentedly on the green grass as they are stalked by an older man with a camcorder.

"I don't know; I get the impression I'm not supposed to really interact with nature here. There's all these railings and signs telling you to stay on the trail." Huckleberry shakes his head.

"Can you eat any of the plants in a national park? Can we make Mormon tea?" Sage asks.

"Technically, yes, but it's not encouraged."

"I feel really constrained here," says Seaweed.

"You know I think it's great, all these old people cruising around in their RVs," Mud joins the conversation. "Except for all the gas they use," she adds. "I mean, at least they aren't sitting at home in front of the TV."

"I saw a couple watching TV in their RV," says Seeker.

"Dude, this is like David Lynch meets the Brady Bunch," Bobofet excitedly articulates the precise metaphor.

"I was just talking to this old lady, and she was great. She was telling me about all the national parks they'd been to and how she and her husband retired and sold their house and just live in their RV traveling around the country. She was really interested in what we were doing. She said she wished she could have done this when she was in college. It reminds me just how lucky we are," Mud continues.

"Dude, what's up with all the sprinklers?" asks Bobofet. "I thought this was a desert."

"It is. The park ranger just told me Capitol Reef hasn't recorded precipitation in eleven months. But remember, it's use it or lose it with water rights. When it became a park, Capitol Reef acquired the senior water rights of Fruita, and if they don't exercise those rights by irrigation, they lose them," I explain.

"Umm, seems like a national park should be keeping the water in the river. Or don't fish count?" says Seeker.

Huckleberry declines the trip to the RV park for hot showers.

"Can I just jump into the creek?" he asks.

"Suit yourself."

The ranger stops by, extracts himself from his air-conditioned patrol car and informs us that Banjo sleeping under the picnic table needs to be tied up, and if we continue to pee on the rocks at the edge of the campground, we will receive a citation.

After working in several national parks, I've begun to wonder about their role in our culture. We revere our national parks. They seem to lie somewhere between the sacred and Disneyland in our national consciousness. Even Wallace Stegner refers to national parks as "playgrounds and shrines." Lacking a national religion or cultural identity, we look to the national parks to meet our spiritual and aesthetic needs, indeed the park service highlights the historical role the parks play in the development of our national identity. It is significant that the parks tend toward monumentalism—the deepest canyon, the tallest mountain, the biggest geyser, the oldest tree, etc. Yet the park service fails us by catering to the most base secularism, homogenizing our experiences into paved roads, tame animals, and crowded campgrounds. A national park has more in common with a feedlot or an assembly line than with a wilderness. Success is measured by the number of visitors served. Instead of reveling in the glory of God, we revel in clean restrooms.

Thoreau warned us about becoming slaves to our possessions. Ironically, in our desire to make nature more accessible, we've barricaded ourselves within our automobiles. We have sold our freedom and spirit for the convenience of the personal auto, which sucks the vitality from our souls.

When I worked at Canyonlands National Park, I once encountered a beautiful mandala someone had made in an inconspicuous alcove. Juniper berries, colored rocks, and swirls in the sand composed the mandala. When I casually mentioned it to my supervisor, she suggested I go back and destroy it. Clearly, a national park is no place for religion, especially an earth-based one.

Across the river from the campground, a wooden boardwalk leads visitors to petroglyphs incised into the sandstone cliff recording the presence of the Fremont Indians, who practiced a truly sustainable existence of small-scale agriculture supplemented by hunting and gathering. A national park that secularizes our experience in nature has supplanted Mormon villagers, whose efforts to transform the landscape into a pastoral vision of England assumed religious importance. Peeling back another layer of history reveals the indigenous peoples made no distinction between the land, their lifestyle, and their religion. Like the Spanish cathedrals of Cuzco built upon the

Bighorn sheep, Pleasant Creek, 2015.

foundations of Inca temples, we find strata of human passages piled atop one another in this one place, each formation revealing a distinct epistemology toward the landscape.

Perhaps the wildlife provide the most glaring example. The Native Americans hunted deer and bighorn through the area for thousands of years. However, within thirty years the Mormon settlers had wiped out all the deer in the valley, and the bighorn had been extirpated because of disease contracted from domestic sheep. Now the park service has begun reintroducing bighorn into Capitol Reef, yet the population remains small and isolated. But the deer have returned with a vengeance. The campground provides abundant, albeit unnatural, food and protection from hunting and predators. (Although mountain lions will occasionally venture down in the winter when the campground is empty.) The result is an artificially high deer population comprised of small and undernourished animals.

In the evening I take my sleeping bag down to the river to sleep. "Is that legal?" Metta asks.

I shrug. "Doesn't matter; no one would ever know. Sleeping outside your designated spot, is outside the realm of their consciousness."

48

# ROCK GLOW, SKY SHINE

## The Spirit of Capitol Reef (1978)

STEPHEN TRIMBLE

*After my season as a park interpreter in 1975, I spent the summer of 1977 photographing for a booklet I wrote for the Capitol Reef Natural History Association. We emphasized the backcountry of the park, hoping to complement existing publications that rarely ventured beyond the Scenic Drive.*

Capitol Reef exists in grand terms—the hundred miles of the Fold, wide panoramas in Cathedral Valley, millions of years of time. But it also leads a day-by-day, nook-and-cranny life. This land shares its intimacies most generously when you walk, gazing up at canyon walls, or simply watching what passes under your feet. The images that flow by—rock, flower, lizard, tracks in the sand—all tell tales full of the land's character.

Canyons may be the best part of Capitol Reef: side canyons that twist their way out of the Fold, channeling floodwaters toward the rivers. Their flood-cleared passageways make walking easy, and you will find them everywhere—from the Visitor Center, where Sulphur Creek emerges after carving its Goosenecks, to the most isolated tributaries of Halls Creek in the south end of the park.

### Paradise and slickrock

Each canyon shelters unique secrets. But to share them you often must start walking in desert flats where only a few hardy shrubs grow. An arroyo leads you on toward the Fold, the dry stream bed winding through hills baking under blinding sunlight. The desert molds the personalities of all that grow here, where aridity rules life. To survive, shrubs like blackbrush do away with water-wasting broadleaved greenery, and conspire to live by appearing dead—coping with drought by concentrating their life in a few prickly branches.

For shade you must retreat to low hills, where soil holds a fraction more moisture. Here, a few scattered and struggling juniper and piñon pine make up the evergreen "forest" of the canyons. No other trees can last through the dry spells of this land. Armored with shaggy bark and waxy needles to protect their inner moistness, they edge roots dozens of feet to reach the moist sand that again allows growth.

Suddenly you leave the flats behind and enter the Fold. Sandstone walls rise on either side to funnel you into a canyon. Amble on and you may find water: a trickle, a pool, a spring—tucked back under an overhang or nudged up against a shady cliff. Sheltered in an alcove, nourished by a seep, a hanging garden—the most distinctive canyon country place—shimmers with the brilliance of blossoming monkeyflowers and columbines. *Green* bounces off canyon walls—the transcendent lime-green of maidenhair fern growing like a natural billboard to announce the seep.

And somehow, this tiny algae-scummed spring has all the mystery of the sea. The slow plop of one droplet at a time, falling from seep to ledge to pool, has the same resonance as waves crashing along the Oregon Coast. For that gentle drip means life. It means cottonwood and box elder and shade, and tracks at the pool's muddy edge. Night visitors—cougar, mule deer, canyon mouse—tramp over daytime tracks of dove, raven, antelope ground squirrel.

The slow dripping of a canyon seep rings deep, whispers "Paradise!" more than the lushest mountain meadow. Contrast with the surrounding land makes the smallest patch of green seem positively exotic, for it is hemmed in on all sides by unadorned canyon walls, by old, overwhelming rock.

So much rock, in fact, that canyons can make you a bit uneasy. Past the spring the canyon enters its narrows. Slickrock cliffs soaring a thousand feet on either side make you feel very small. Shadows cool your skin—wet with sweat from the open hotbox of the canyon behind you. The narrows close to a slot barely wide enough to inch through sideways, where full sunlight never reaches. You shinny over a boulder, warily looking thirty feet above you to a logjam left by the last flood. Such places make useless your complacent reactions to the everyday world.

You find yourself nervously seeking kindred souls—a juniper, a lizard—anything other than the ancient rock, anything *alive.* The power of the rocks noticeably humbles the living things scattered among them. Yet suddenly you sense *life* flowing through a rigid claw of juniper. A lizard's flash from under a barberry bush startles the cliffs with movement, sends an irrepressible cry echoing down the canyon: "I'm ALIVE. If only for a moment, I race across cliffs, my claws scratch rock, I mate and reproduce my kind. I give this land its reality, not rock." Life goes on, if only to spite the watching cliffs.

You hear the lizard's scuttle and can't help but breathe easier, reminded that you are not alone with the rock. You move along the canyon floor once more, after a sheepish look over your shoulder at impassive canyon walls.

### The wild heart of the fold

Along your walk, you look the Reef in the eye, and find there a world of images. Canyon walls provide both frame and picture, tapestried by dark mineral stains of desert varnish, accented at their bases by twisted juniper

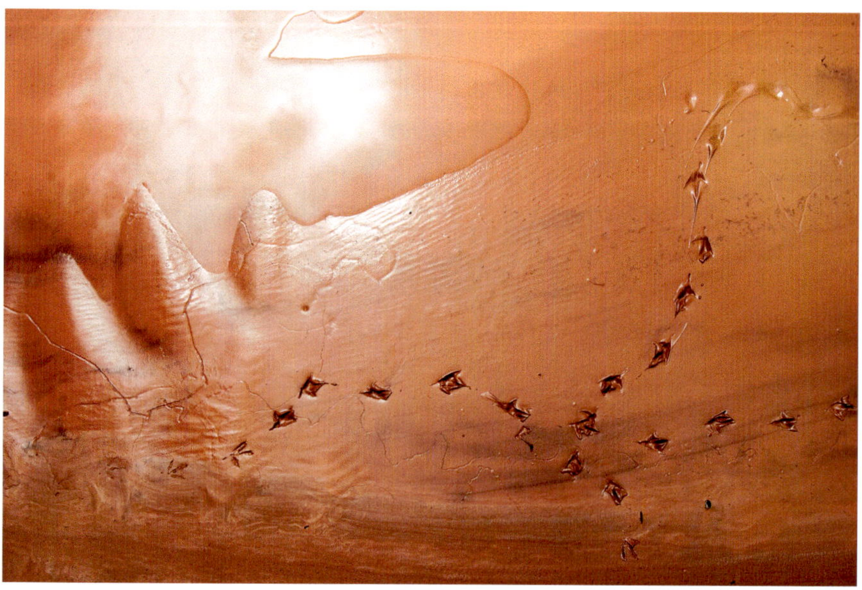

Tracks, Sulphur Creek, 2007.

gnarls and murky plunge pools. Claret-cup cactus blooms against all odds from a crack in a twenty-ton boulder. Leopard frogs shine velvet-gold in the morning sun. And canyon wrens sing—the lilting trill of a musical waterfall, the theme song of the canyon country.

Past the narrows a trickle of water leads you on. Here and there a single fallen cottonwood leaf gleams through golden riffles of the stream. Another small narrows opens up into an amphitheater—a cool grotto of stone in the secret center of the Fold.

Soak your feet here in the still water of a plunge pool, tiny water insects brushing your legs. The canyon continues on above the amphitheater, but a thirty-foot dryfall blocks the way. There is a chance no one has explored the upper reaches of this canyon.

The essence of Capitol Reef waits above that amphitheater, where perhaps no person has walked. For the spirit of this place is wildness, always waiting around an unexplored bend.

When you return to Capitol Reef, walk any canyon and soothe yourself with the bright music of canyon wrens. Squeeze through a narrows, sit yourself down in a plunge pool. But leave one rimfall unclimbed, one bend unexplored. Just once, stand and glory in the magic luring you around a corner, and then turn back the way you came. Leave one canyon to wind its way around a cliff and disappear unseen into the unknown, wild heart of the Fold.

# 49

## OCTOBER JOURNEY (1996)

ROSE HOUK

*Flagstaff writer Rose Houk's lifetime of books about national parks includes two about Capitol Reef.* Canyon Country Eden *captures the spirit of this place in personal narratives like this excerpt, a walk down Halls Creek.*

Near an old oil drillers' site on Big Thompson Mesa, a faint path drops into Halls Creek. Cairns mark the route, the only break in the cliffs for miles in either direction. I look over the edge and gulp hard. I've been this way before.

Starting down, my knees balk at what they're asked to do. I clutch the branches of shrubs for support, bolstered by words of encouragement from my husband, Michael. Our descent goes well, and we soon enter the wide valley of Halls Creek.

We pause under a lone cottonwood dressed in October gold. From here, Halls Creek heads due south, flanked by the soaring swell of the Waterpocket Fold to the west and by the impenetrable cliffs of Halls Mesa to the east. This country is the remotest of the remote in Capitol Reef.

We could follow the meandering, moist creekbed, stepping from boulder to boulder, or take a shortcut across the flats along an old wagon road. We choose the shortcut, a sandy trail first broken by horse hooves, boot heels, and wooden wheels. The path is fringed by sagebrush, Indian ricegrass, and prickly pear adorned with ripe red fruit. One of the cactus fruits lies near the entrance to a burrow where ants carry away its seeds.

After a couple of miles, I shed my pack and seek the irresistible shade of another cottonwood. The only sound is the rattling of the burnished leaves in the breeze. A woodpecker flies to a branch in the tree. From a nearby hillock, Michael calls me over to see the translucent molted skin of a snake twined through a rabbitbrush.

As we walk the old road, I think about the journey of Mormon pioneers through Halls Creek, also in October, but in 1882. Leading the way was Charles Hall, for whom this creek is named. A native of Maine, a carpenter, bricklayer, and boatbuilder, Hall came to Utah in 1850. In the winter of 1879–1880, he was part of the epic Hole-in-the-Rock Expedition, in which a group of Latter-day Saints lowered wagons and stock two thousand feet down a precipitous break in a cliff to reach the Colorado River. Hall and his sons ferried the Saints across the river, and the pioneers continued on to the San Juan River, where they settled the town of Bluff.

Soon after this episode, Charles Hall began his search for a more accessible route to the Colorado River. He blazed a trail east from Escalante, over the Waterpocket Fold, and south down Halls Creek to its junction with the river. Here, he built a wooden ferry, ten feet by thirty feet, and went into business. He charged his customers five dollars a wagon and seventy-five cents a horse to carry them across the Colorado's swift current.

In late October 1882, a small wagon train of Saints set out to join the settlers at Bluff. They followed Hall's new "road" through Harris Wash, up Silver Falls Creek and across the Circle Cliffs, through Muley Twist Canyon, then down Halls Creek to Halls Crossing. Among them was Josephine Wood, known to all as "Aunt Jody." Her diary entries vividly recount their travails. "Started...with great sorrow and weeping because of parting from all of our dear friends and relatives...we had nothing to do but to lie back in our wagons and think of those we had left behind." As they moved overland to the Waterpocket Fold, blowing wind, scarce water, and rough country became the order of each grueling day. The women and children often had to push the wagons through knee-deep sand. At one point Aunt Jody declared: "It is the most God-forsaken and wild looking country that was ever traveled." Their salvation was the waterpockets, which still held rainwater from late-summer cloudbursts.

Halls Creek Narrows, 1977.

The Saints finally arrived at Halls Crossing on November 3, where they had to face one more ordeal. "Now it is our turn—O pray for us!" beseeched Aunt Jody. She and her children got inside their wagon, which was then driven onto the raft and lashed down. "The men started rowing, and down the raft and all went into the water with a splash. My heart went faint, I went blind and clung to my babies. I shall never forget my feeling as we went down in to the water, and my fear of the wagon going off into the swift-flowing river." But they survived the crossing, and once they were safe on the opposite bank Aunt Jody and her companions "did thank our Heavenly Father."

Our trek down Halls Creek is not nearly so eventful. We sidestep slow-moving pools of water and try to avoid the slippery gray mud left by a recent flood. The water and mud are reminders of the paradox of this place: having too much water or scarcely enough.

One summer afternoon, while on another trip into the canyons of Halls Creek, dark clouds gathered over the Fold and we decided to pack up and make a hasty departure. When we were halfway up the trail, the storm swept in.

We took refuge under a huge overhanging boulder. Lightning snapped in the void before us and waterfalls poured over the Fold. As we stared wide-eyed, Halls Creek started to flow. Everything I'd read about desert flash floods had not prepared me for the sight of the proverbial wall of water pouring downstream like molten copper, bank to bank, blocking access and carelessly plucking up anything in its path.

From the supposed safety of our boulder shelter, we watched this drama for a long time, happy to be high and dry above Halls Creek. Then, the trail in front of us turned into a stream of water.

Michael suggested that if the rock we were sitting under showed any sign of movement, we should hightail it out, storm or no storm. The thought had not even crossed my mind, but I heeded the warning. A geologist by training, he knew rocks can move. The boulder never budged, and when the storm ceased we trudged out to the mesa top.

I recall this adventure as we approach Halls Narrows, our destination for the night. The Narrows is a three-mile stretch of tight twists and turns, a delectable detour before Halls Creek resumes its final, no-nonsense course southward to Halls Crossing, now submerged beneath Lake Powell.

Just when our bodies begin to weary from the day's journey, the Narrows come into view. We unpack our gear beneath a great vaulted alcove of Navajo Sandstone, a chamber that could easily seat a symphony orchestra. The walls are 180 degrees above us. The creek, rippling softly in the background, is lined with water-loving willows, oaks, boxelders, barberries, and cottonwoods. A hanging garden of brilliant emerald maidenhair ferns drapes a sandstone cleft.

As we start supper, the sharp crack of falling rock leads us to examine the boulders on the sand around us. We look up. It's all too clear where they've come from. Just as we're trying to decide whether we've chosen the best spot, a small chunk of rock clunks onto our sleeping pads. Deliberations cease. We pick up everything and move a short distance back upstream to an open, sandy site.

Bats flutter by in the twilight. The first stars shimmer in the lens of sky overhead. As dark flows in, the cloud of the Milky Way follows the curve of the rock above us. Delphinus, the Dolphin, leaps through the sky: summer stars changing places with winter stars. The moon rises, just past full, and washes the walls with silver. It is a still, still night, filled with dreams.

Near dawn, a canyon wren sings us awake, and a sliver of morning sunlight slides down the alcove wall. We watch to see whether the light will reach the canyon bottom. It doesn't. After a cup of coffee, I wrest myself from the cocoon of my sleeping bag, looking forward to a day dawdling through the Narrows, going as far as we wish, without backpacks. Ravens fly over. One squawks just to hear itself squawk, or perhaps to urge me to get going so it can investigate my pack.

We filter water, fill bottles, stuff a day pack with lunch, and start walking. Our progress is slow because there's so much to examine: the velvety yellow leaves of the boxelders, their twinned key fruits hanging in pendulous chains; a strange brown mushroom sprouting in the wet sand; stiff scouring rush flattened by the last pulse of floodwater; a hefty chunk of petrified wood brought downstream from some distant point; a tiny canyon tree frog.

Slippery mud waits to suck us in as we step carefully across it. In pools where I can't see bottom, I wade in and sink down another foot, slip, and go in up to my waist. To dry out, I perch on a sunny boulder and pour the thick silt from my boots. I spot a small alcove hidden by a thicket of Gambel oak and go up to investigate. Poised on the alcove's sandy floor is the skull of a bighorn ram.

With each new bend of the creek, the view is more stunning. I round a corner and gaze downstream at a glowing, backlit cottonwood that stands in perfect composition. It's an image forever preserved in my memory, like a wildflower pressed between the pages of a book. Perhaps that sight brings on a late-afternoon melancholy. As I sit by the stream sipping cool water from my metal cup, I sense the hope and promise of morning sliding away, replaced with an urge to go back, settle in for the night, cook dinner, stay warm. I know this feeling. It's a conflict between utter happiness and contentment in this wild, unpeopled place and my human desire for a secure nest.

We wind back through the Narrows, making camp on a generous, flat spit of sand. High, thin clouds float over the Fold, but evaporate in the dusk. We see more stars and more sky than we did the night before. There are more constellations to name, more stories to tell. It's a long night with a bright moon that never sets. At dawn, pink mare's tails fringe the top of Halls Mesa. A jubilant morning chorus of coyotes spills out of the hills. Frost decorates the sleeping bags. We linger in camp, cleaning up, washing breakfast dishes, packing away damp, dirty clothes. We start back in the coolness, passing now-familiar landmarks.

We came to Halls Creek to watch and to listen. We have seen no one during our three days here. We didn't expect to. How different it must have been for Josephine Wood and the other settlers of this country, who came not to recreate, but to try to wrest a living from this hard country. Like us, were they sometimes overwhelmed by the beauty and the silence of this magnificent landscape? Did they find solace? I don't know, but I hope so.

# 50

## THE TROUBLE WITH CELL PHONES (2008)

CRAIG CHILDS

*In more than a dozen books, Colorado writer Craig Childs explores "the relationship between humans, animals, landscape, and time," mostly in the American deserts. He tells us that he "travels the interstitial places. Roads diverge in the wood and I start climbing trees." This piece comes from his website archive of writing about places.*

There are many things I said I would not do, and I've been steadily checking them off my list, the various calamities of fatherhood, the small television upstairs, the exercise machine. Tonight's came by surprise. I pulled the truck over at dusk at the edge of Capitol Reef in southern Utah, taking a dirt road out to a place to sleep, pack ready to go in the bed. I parked, shouldered the pack, and walked into an indigo sheen of horizon. Longing for Regan and the children, I brought the cell phone along, surprised to find coverage out here. Is it so wrong to have a cell phone? I know my vows. I don't have to be reminded. I don't take cell phones into the wilderness. We mesh our lives with safety nets, people calling from mountain tops pleading for a helicopter to come get them. We wander around with tethers, reel ourselves back in the moment it gets messy. It is as if we are not wholly there, a pause button ready to halt our experience. People even give me crap for not carrying a cell phone out there. They think it's wrong not to, somehow irresponsible. It's bad enough I've got one at all. I'd rather not be rescued. But for christsake this was not wilderness. I could see headlights, lonely dopplers driving to

Hiker at Twin Rocks, 2015.

Hanksville. I walked toward the strange hats of sandstone up against the sky, pulled out the phone as I went, speed-dialed home. You'll never guess what I'm doing! I'm walking around with my pack calling you! They screamed, laughed. I dropped my pack and laid against it, found the north star and imagined the view from home. What I won't do for love. My heart aside for the moment, it felt much the way I thought it would, the reason I vowed not to do it. I was suddenly somewhere else. It felt like a layer of Plexiglas went up, even between me and the stars, which had become numerous since we started talking. I was simultaneously here and not, my reach digitally altered so that I could cover the globe without knowing a single god damned route, just fingers punching buttons. Even writing in a journal out here has often put me behind a veil thin as a blade of grass, through which I write to you, unexpected reader. You are not here, sitting on sandstone with me, yet I am talking to you, grabbing hold of a rope that leads to a hole where I pull myself out of this place, off of this smooth deck of bedrock where now I say good night to my family, turn off the phone, and write this note to you in my journal. I settle to sleep in a cup of stone, my still body pressed against the sky.

# 51

## DISCOVER YOUR OWN PRIVATE PARADISE (2011)

STEVE HOWE

*Founder of Redrock Adventure Guides and field editor for* Backpacker *magazine, Steve Howe has been backpacking, climbing, kayaking, skiing, and trail running for forty-five years. He's been based in Torrey since 1990. "Capitol Reef and Dirty Devil country are the reasons I'm here," he says. "This is one of the finest wilderness expanses left on earth, and very few people truly see it."*

I thought the rope would solve my problem, but as can happen with ropes, it only led me deeper into trouble. I was trying to extend an unproven route through Capitol Reef, solo, and the rope had enabled me to descend one more rotten cliff band. But rigging an anchor in loose boulders had eaten up time. Now I was standing on an isolated sandstone fin that protruded out over a yawning canyon, with no way forward. Dead-ended again.

Fortunately, I'd long since learned that getting stymied in the Waterpocket Fold was never a true failure. I knew that finding which routes didn't work would eventually help me zero in on ones that did. And I'd discovered yet another site to put on my personal, expanding map of Capitol Reef. The table-flat mezzanine that had thwarted me was easily the most spectacular campsite I'd ever stumbled across. To my east was a poster view of the snow-capped Henry Mountains, rising like abandoned pyramids. To the west, 1,000-foot cliffs of blonde sandstone slammed in close, their walls like overlapping El Capitans, interweaving to pinch off the gorge's headwaters. But I had no overnight gear, my water was long gone, and the late October temperature was quickly skydiving from warm to frigid. So as sunset dimmed toward darkness, I dug my headlamp out, rigged my ascenders, and began the long, black retreat toward home with a bittersweet pang of regret. I had not succeeded, but I had not failed. I'd punched the route a few more miles. And I'd be back.

It was a cold January in 1990 when I decided, on the spot, to make Capitol Reef National Park my backyard. A friend and I had backpacked into the deep, ochre cleft of Spring Canyon, post-holing down the sandstone gorge through month-old snow that didn't have a single footprint. We camped in a narrow sandstone hallway underneath a corkscrew pinnacle that reared defiantly overhead like it was giving the finger to the gods of erosion. We tucked deep into our bags as the stars came out and the canyon became a sandstone freezer. Around midnight, I crawled out of the tent into a scene painted by glowing moonlight, so still my heartbeat sounded like war drums.

I'd visited Capitol Reef before, in warmer seasons, and been staggered by the audacious geology of the Waterpocket Fold, a brick-red Wingate escarpment topped by whimsical cones of blonde Navajo sandstone. But it wasn't slickrock scenery alone that compelled me to move to the nearby hamlet of Torrey. Capitol Reef is special in other ways. There are only a handful of developed trails. Most of the landmarks remain unnamed. The roadside attractions are so-so. The wildest topography lies concealed, accessible only by tough foot travel. In order to find Capitol Reef's best places, you have to follow your curiosity through crisscrossed gullies, over anonymous slickrock saddles, and across crumbling boulder fields. You almost never reach your goal on the first try. My kind of place.

I spent untold hours running along sandy washes and countless days scrambling through the deep, meandering canyons of my new backyard, often returning by the dim glow of a headlamp. Atop the tortured geology of the Waterpocket Fold, in this place Ute Indians named "land of the sleeping rainbow," I discovered tans and oranges and shades of vermilion I'd never seen before. And I found my own personal paradise.

The routefinding was more complex than anything I'd ever encountered. Gorges pinched off into dead-ends or died in pouroffs. Cliff bands hid between the contour lines of topo maps. Even aerial photos didn't have enough detail to resolve the maze of crag and canyon, ledge and gully.

Simple out-and-back hikes often turned into mini epics. I learned to rely on obscure signposts like odd-shaped trees and individual boulders. Each trip felt so much more personal, and wild, than it would if I were following signed trails. I reveled in the childlike joys of real exploration. And the more I probed, the more I wanted to find a route through the northern Waterpocket Fold's soaring domes and faulted canyons—in one side, out the other, witnessing all the geologic mysteries that lay in between.

Capitol Reef is a long, linear park, running 100 miles north to south from the badlands of Cathedral Valley to the narrow canyons of Halls Creek and the Muley Twists. In the south, it's defined by a long, narrow series of tilted sandstone flatirons, in places less than a mile across. But I was always drawn to the northern reaches, where the Fold is thicker, twisted and broken into a jumble of weathered sandstone peaks crisscrossed by knife-straight faults. It's a dense desert mountain range, much of it impassable—unless you like to climb dangerously soft rock with zero protection. Slowly, I became obsessed with finding safe ways through the forbidding maze.

Had others done so before me? Maybe. Slickrock country is, after all, the kingdom of hermits, and people here treasure their supposed secrets to an amusing degree. But the more I advanced, the more I came to believe that maybe no one had ever wandered exactly here. There were no local legends, no mapped routes, no desert rat beta, no cairns or paths or even broken branches. The only prints I ever saw were cougar, coyote, and deer, later joined by desert bighorn sheep once they were reintroduced in 1996. Even now, 20 years later, I've never seen another hiker off-trail.

I became acutely aware of the consequences of a misstep, while alone, out in some obscure side canyon. Rocks roll, ledges crumble, branches break. I took to hauling bivouac gear, even on short jaunts and trail runs. While most places get tamer over time, Capitol Reef just got bigger and badder.

The first time I actually tried thru-hiking the 17-mile section across the Fold, between Capitol Gorge and Grand Wash, was in 1994, with four friends. We were all lifelong outdoor pros, but we only made it nine miles in three days before retreating down narrow ledges and sketchy sandstone slabs, headlamp batteries dying, tails between our legs. The next year, I managed to pull it off with a pair of New Zealanders, though certainly not by the best way. We spent most of our time in overgrown gullies, and used a rope repeatedly. The success left me surprisingly unfulfilled. It was an inelegant and risky route, stupid even. I wanted a more scenic, less technical path, with room to wander.

That search eventually turned into a 10-year, stop-and-go effort. Once GPS units and desktop mapping improved, I could explore and later see exactly where I had been, and how that related to neighboring peaks or gullies. It soon became clear I wasn't looking for a single route, but a braided series of possibilities, weaving the best threads together into an original fabric.

With this revelation, it became less about finding a way through and more about finding hidden gems within. A ponderosa-ringed pothole here, a gymnasium-size slickrock bowl there, a pool of tadpoles, a vertigo-inducing overlook, or a bighorn trail through ancient, fragile cryptobiotic

South along the top of the Fold, Deep Creek, 2012.

soil [*the slow-growing erosion-blocking living soil crust, easily destroyed by a careless footstep or vehicle track*]. I finally put all the links together and completed the point-to-point journey on a 2005 solo. But by then I knew I wasn't done. The finish line changes. I'm still working on extending the route farther north and south, finding ever better campsites, imaginative detours, and hidden passes.

The best part? After all those years of exploration, I'm sharing my secret hideaway—yes, including all of that hard-earned GPS data—with you. But make no mistake: Even when you know the way, this backcountry demands respect. I've completed the traverse half a dozen times, and I still have to bail when conditions are bad, or improvise detours because of ice-slickened rock or flood-caused changes.

Of course, I want you to exercise good LNT [*Leave No Trace*] etiquette, but I'm not too worried about newcomers messing up my backyard. Most of the route is on slickrock or sandy wash. With a little smarts, you can travel the entire way leaving literally no mark. Stay off the cryptobiotic soil, and follow game trails when you must cross small patches. Don't build fires because they're illegal. And don't leave cairns or campsite rock piles because that spoils the adventure for the rest of us.

So come see why Capitol Reef is the national park system's best-kept secret. Or better yet, discover your own private paradise here or in any other wilderness. I've found equally remarkable pockets of wild terrain in some of America's busiest parks—deserted passes in Yosemite, vacant slickrock plateaus in Zion, secret rhododendron-lined swimming holes in Great Smoky Mountains. You can do the same: Just develop the skills and confidence to get off the trail, and forge your own route. Capitol Reef has taught me that anyone can discover a whole new level of adventure, wonder, and commitment simply by looking at a pass, a ridge, or a blank spot on the trail map, and wondering, "Can I get to the other side…?" Do it.

# 52

## SPOKED DREAMS

### An Odyssey by Bicycle and Mind (2005)

CHARLES RIDDEL

*Texas writer and engineer Charles Riddel's 2005* Spoked Dreams *chronicles a 2,500-mile solo bicycle trip through six southwestern states. As Riddel says, "the real story is about bicycling from one place in life to another." In this excerpt from near the beginning of the journey, Riddel arrives in Capitol Reef from Salt Lake City.*

### Day 5:

I believe there is a child-like dimension in people that mostly lies dormant. An encounter with a crazy guy on a bike in darn near the middle of nowhere, challenging himself and nature in such a simple manner, seems to liberate the dormant child in the people I meet, at least for a few wistful moments. Some folks ask me questions as if they are starting their own bicycle trip tomorrow, and then they climb back into their motorhomes and drive towards Lake Powell. I've chatted with a number of retirees on extended escapes. Many have bikes lashed to the back of their motorhomes. Mostly, they use the bikes to get around campgrounds, but I can tell from the questions they ask that many of them still like to dream young.

I coasted to a stop in front of the visitor center at Capitol Reef National Park in mid-afternoon. The day turned a bit cloudy, providing some welcome relief as I completed the ride. The tide of kindness from strangers continued to rise when the rangers declared that someone who had biked so far must camp for free.

I rode the short distance to the campground and selected a site....

This is my first real campsite, and I am clearly enjoying it. The amenities and security of the campground feel quite luxurious to me. But I am feeling something else, too—a sense of disbelief that I actually made it here, all the way from Salt Lake City on a bicycle. I also feel a bit of vulnerability about being in a remote setting with only a bike to carry me home—wherever home really is, now.

Not far up the campground loop is a site harboring two large, identical motorhomes and two jeeps that were towed here behind the motorhomes. I dub this site "nesting motorhomes with offspring." The contrast between my little campsite and the one with the nesting pair of motorhomes underscores my sense of displacement. But it also strengthens my child-like fascination with making this experiment work. I can visualize myself being eight years old again, standing at the edge of old US Route 80 [*in Arlington, Texas*] wondering if I can see any mountains by staring hard enough into the sunset, where its blacktop lanes converge....

I finish my feast, and tire of idle musings and sitting still. I return to the visitor center for a while and look at the old schoolhouse. What would it have been like to go to school in this one-room schoolhouse? Would the rimming canyon walls and tiny schoolhouse create a sense of isolation, like a group of kids stranded in a lifeboat waiting for rescue, listening to repetitive stories to pass the hours? Did the students sometimes hike to the highest viewpoint to discuss the grand sweep of the rumpled terrain and talk about the wonders crouching beyond the horizon in various directions? Was this school beyond the reach of new ideas and discoveries, or did new ideas dance spontaneously amid the mixed chatter of this shoebox schoolhouse?

I am trying to be in that time rather than looking backward to that time. It's not easy to imagine "then" without the prejudices of "now." What is more mystifying: to look at the world when essentially everything is a mystery but the true complexities are not suspected—or to look at the world when the complicated workings are more appreciated but the underlying reasons are still not understood?...

Before sunset, I take a short hike along the Fremont River and then return to camp. Around the tabletop in my luxury-suite campsite, the day dissolves into darkness. I play with my long shadows thrown down by the lanterns dotting the campground. I brush my teeth. I stow my gear for the night and scoot backwards into my little tent. Already, solitude is the heaviest load I carry.

Total mileage for the day: 45.5

### Day 6:

My eyes blink open with the sky already turning from gray to blue but with no sunlight yet varnishing the sides of the nearby mountains. Today is a hiking day, no biking at all. The targeted hike extends from the campground to the high point on the Frying Pan Trail. The campground is at an elevation of about 5,400 feet, and my turn-around at Lookout Point will be above 6,400 feet.

Capitol Reef is a land of soft rock and hard elements. The countless thrusts of nature's tiny swords have whittled deep wounds into the rainbow entrails of the land. But many of the same elements have made, with a little help from man in some cases, the very softest parts of the land green and fertile.

The earth here is not just dissected; it is sculpted. The complexity of form is as if the rock were suddenly poured from the sky and frozen in place at mid-splash. The enormous foundations of the convoluted crests tilt upward into great, rusted plunges. The long, rutted cliffs sometimes stretch to the horizon. While the valleys and hillsides are liberally (at least for a desert) covered with vegetation, the immense cliffs are scrubbed barren. In this place, the outer limits of visible life are clearly defined.

Capitol Reef is a focal point for the staggering sweep of geologic time and the crushing forces buried in the earth. The intensity of The Reef stems from its piggy-back ride along the spine of a hundred-mile long crease in the earth, called the Waterpocket Fold, that was pinched upward as the immense and lofty Colorado Plateau was heaved into the sky, far beyond the horizon to the east.

Wherever great cliffs are seen, huge amounts of time are exposed.

Above Cohab Canyon on the Frying Pan Trail, 2004.

The eroded flanks of The Reef bare a lengthy cross-section of Earth's history dating backward to the Permian Period, 250 million years into the dimness of the past. But Capitol Reef is unusual in that time is stretched across its horizontal floors as well as its vertical walls. This is because of its position on the crest of the great fold. As the layers of rock were buckled and bent upward over the past ten million years, the folded edges have been clawed away by the erosive forces of the earth—first, along a narrow line in the youngest and topmost layers of rock, then spreading outward east and west away from the axis of the fold as the bending and thrusting continued—ever wider in space and ever deeper in time. The tilted edges of the rock layers jut above the fringing valleys because the upward thrust slightly outpaces the downward carving. The oldest formations straddle the center of the fold, with the very oldest rocks bleeding through the incisions at the bottoms of the canyons, near the centerline of the eroded kink....

The top of the trail arrives quickly, three and a half miles in just over an hour. Some rain clouds meander overhead, occasionally thumping out a half-hearted drumbeat of thunder. A light sprinkle teases but does not quench. There is no real threat, and I have plenty of time to linger on the summit.

The vista from this high perch demands inspection. I'm sure that I can see at least seventy-five miles in some directions. Off just north of east, slashes of canyons crisscross a high plateau. This is in the general direction of Canyonlands, a realm crouching within my near future. I strain to see the transitions from here to there. I can distinguish the forms of dissected mesas and see pinkish-red hues reminiscent of Canyonlands and its outer perimeter. But the curvature of the earth, the northern slopes of the Henry Mountains, and the intervening plateaus hide the features of Canyonlands from my line of sight. In my mind's eye only, I see the spires and buttes, the prominent gorges, and the broad basins of Canyonlands. I know they are out there. My compass points the direction.

I also look towards the southeast, in the direction of Lake Powell, another place in my future. The high ground near the lake combines with the deep depression in which the lake sits to conceal it from view. I imagine my route from here to there. It's a long way.

The sightlines are pulled taut by the expanses. I sense magic in the multiplication that my bicycle performs—multiplying the turns of my feet into distances too far to see their other ends. My bike looks so modest. Would anyone else see the Multiplication Magician hidden in the tubes, sprockets and spokes? I am reminded of seeing one of John Wesley Powell's old dories used to explore the Green and Colorado rivers. Looking at it didn't convey the distances covered and the adventures created in it.

The future fades into the hazy distance and the uncertainty of events yet to be. For the moment, I will ride my daydream.

Summer thunderstorm over the Waterpocket Fold from Notom Road, 2017.

# PART IX

## PRACTICABLE RIDGES

### Nurturing the Future

I end with the concluding words in my 2008 book, *Bargaining for Eden*. In this book, I ask: How do we make decisions together about land we hold dear?

In exploring the answers to this question, I track down the motives that inspire passion in two iconic places in the New West: a northern Utah mountain and its historic ski area and a redrock mesa between Torrey and Capitol Reef where my family builds a home. On the mountain, citizens fight to hold on to their sense of community as public land becomes private and the mountain develops on a grand scale prior to hosting the 2002 Salt Lake City Winter Olympics.

In the second story, I become the lens, turning from observer to actor. After buying land within sight of Capitol Reef, I look at Wayne County for examples of both how we remain stuck in our trenches and how we might transcend these unproductive narratives. When my family splits the property, becoming land developers ourselves on a tiny scale, I must deal with the ethics of ownership and learn how to be not just an urban pilgrim visiting the public lands but a responsible tax-paying citizen of Wayne County.

The challenge of Capitol Reef's "Impracticable Ridges" overwhelmed young mapmaker Lieutenant George Wheeler in the 1870s. But we've learned a lot about this hard country since Wheeler, and today's National Park Service must find a "practicable" path that balances our desires and our responsibilities in Capitol Reef. I close *Bargaining for Eden* and this anthology with a manifesto, a Credo that summarizes my hopes for the future of Capitol Reef and for the American West.

# 53

## BARGAINING FOR EDEN (2008)

STEPHEN TRIMBLE

Wayne County life is a classic small-town, long-winter, droughty-summer western challenge. Ambivalence is born and bred; every native navigates between love for the beauty and solitude of the place and resentment of its isolation and lack of economic opportunity. County Commissioner Allen Jones tells me, "My dad sent me off into the red rocks to work, not to play. When I came back after many years, I saw this land as beautiful for the first time."

This place of red rocks and hard work preserves the Big Empties for our imaginations. The nearest stoplight is still two counties and more than sixty miles away in Richfield. About 2,500 people live in Wayne County, scattered across 2,500 square miles—just one person per square mile in an area more than eight times larger than the five boroughs of New York City, which could absorb the entire population of Wayne County in less than ten Manhattan city blocks.

Homogeneity rules. The county is 97.3 percent white. Two-thirds of Utahns are Mormon; in Wayne County 80 percent belong to the Church.

Today, at the beginning of the twenty-first century, no more than five families in Wayne County are able to live on their farm or ranch earnings alone—and everybody knows who they are. More families in the county depend on earnings from art than from agriculture. Most wage-earners need to supplement the microprofits from raising alfalfa and running a few cows, sheep, or the occasional pair of llamas. Paychecks come from many other sources, including waitressing, carpentering, teaching and permanent or

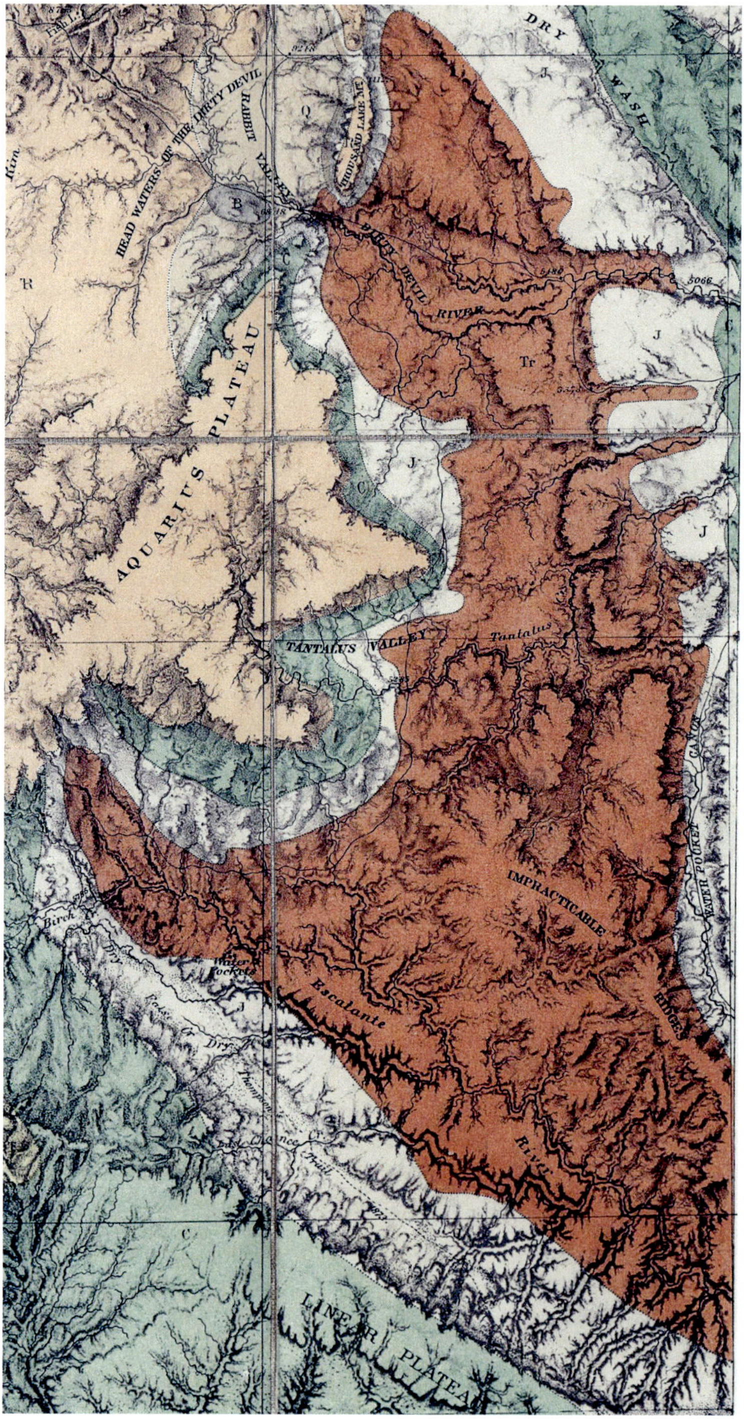


Chuck Bowden says of George Wheeler's map of southern Utah, the cartography "captures the dust and heat and thirst of the expeditions of 1872 and 1873 in fine lines, reds, greens, oranges, browns, yellows, and blues" (see Chapter 16 of this book). In the lower right corner, Wheeler surrenders to the difficulty of Capitol Reef's "Impracticable Ridges."

seasonal work with government agencies. In acreage farmed, Wayne County ranks third to last in Utah, a testament to the minimal amount of arable land.

There are three clusters of residences in the Torrey Valley, where the Fremont River flows between the 11,000-foot-high plateaus of Thousand Lake Mountain and Boulder Mountain. Teasdale is tiny; Grover—with perhaps a dozen full-time residents living in dispersed ranches and cabins where the scenic highway heads south, preparing to lilt up into the ponderosas and aspen of Boulder Mountain—even tinier. Torrey is the valley's metropolis and only incorporated town, 170 citizens living in a grid four blocks by six blocks along the all-important irrigation ditch that parallels Main Street. Torrey is a crossroads, where the route coming north from Bryce Canyon and Grand Staircase meets the east-west highway bisecting Capitol Reef National Park just four miles to the east.

Torrey has a fine restaurant, whose chef trained at Deer Valley Resort and serves organic veggies and good crusty bread from a back-to-the-land farmer who is proud of introducing his rancher neighbors to *bruschetta.* Galleries and bookstores are proliferating in town. This is gentrification, with all of its trade-offs.

This gateway town for Capitol Reef will never boom beyond imagining, however. It can't. Fred and Janet Hansen, who have taken turns serving as mayor ever since retiring here from northern Utah twenty-five years ago, worry about the Torrey water supply, which depends on mountain springs. Four years after we hooked up our meter in 2000, the town hit its limit, and currently no new home can connect to the city water systems unless it has an existing meter. Fred has been mayor in recent years: "I'm sorry this drought caught up with us while I'm mayor, but it's good for the community. It woke us up. We just quit selling water; we can't sell something we don't have. That will determine the limits of our growth. Maybe that's why God set up the water situation—to let us know where the limits of the desert lie."

I first lived in Wayne County thirty years ago when I worked as a Capitol Reef ranger; I've visited nearly every year since. I've met three generations of local residents. I don't feel like a newcomer, because I've loved this country for so long. But newcomer I am.

The rural sociologists who study "natural amenity-based growth" in the West apply scholarly analysis to the issues in this book. They support my

Fourth of July parade, Torrey, 2007.

conviction that if we can puzzle out a way to talk with each other we will find overlapping values that can lead us to higher ground; we can reach islands in the seething sea of antipathy, and from there, as a community, we can set a course for the future and stand strong against the grand development schemes. . . .

Old-timers and newcomers—the "amenity migrants" who have lived in rural communities for five years or less—show similar levels of concern for the environment. Newcomers bond first with the landscape, reinforcing the "community of place." Old-timers are more invested in social networks, the "community of interests"; in rural Utah, membership in the LDS church guarantees instant connection. But the expected culture clash doesn't pop out from statistics. Strong opinions exist at the ends of the curve; in the middle, communities harbor common ground.

One friend who moved to Wayne County from the Northeast, in love with the enchanted wildness of the redrock landscape, found herself stranded at midlife in an isolated place that she wasn't sure she had actively chosen, living behind what one local calls the Alfalfa Curtain. At the end of one long winter she admitted that she had become obsessed with the question "How do you mitigate regret?"

Our whole community needs to address this question before we have taken too many actions that might require mitigation, before we approve too many steps toward homogeneity that will be cause for regret. I can feel the possibilities out there if we avoid these mistakes; I can picture a future filled with thoughtful dialogue, tantalizing us with hope.

Rural Utah has always been a culture of villages, but more than three hundred homes now lie scattered outside the three Torrey valley communities. Noisy newcomer WOES [*deemed "Wealthy Over-Educated Spoiled brats" by one longtime local*] have been growing in numbers and shifting the culture. I can join the Wayne County community now without feeling quite so conspicuous. Indeed, one progressive grassroots land-use coalition—the Friends of Grover—has been working on threats to public lands in Wayne County since the mid-1990s, and I was pleased to sign on as a member.

Don't let me make too much of this. I perch on the rim of rural life without grappling with its most difficult daily realities. But I long for a chance to cooperate, for the satisfaction of protecting and ensuring a future that the community imagines together. I hope to work from what some call the Radical Center, the place where all factions come to talk. If we can move beyond our trenches, I'm convinced that we can find shared values in our mutual opposition to distant corporate and political powers refusing to listen to the people who cherish the land.

...As I began to imagine sitting across the big table from people who are feeling embattled and defensive, I thought about how best to distill my principles before blundering ahead. I tried to list what I believe to be true, an inventory of issues that newcomers and old-timers alike must grapple with and within which we must find bridges to take us to a sustainable future together.

I'm not the first to attempt to articulate the bedrock of my creed. Victor Scheffer, the eminent marine biologist, stated environmentalism's "Articles of Faith" a few years ago. To me Scheffer's words seem like simple declarations of truth.

1. All things are connected.
2. Earthly goods are limited.
3. Nature's way is best.
4. The survival of humankind depends on natural diversity.
5. Environmentalism is radical.

These are radical notions. They demand fundamental change—in political systems, law, agriculture and industry, the structure of capitalism, internationalism, and education. They constitute, as Scheffer acknowledges, "a morality of life or death for civilization."

I first encountered Scheffer's manifesto on an anti-environmental "wise use" web site, where it was cited as a missive from the enemy, a distillation of "eco-ideology"—and that word was always best said with a sneer. The wise use countermanifesto places humans at the center of the natural world, with all other beings servicing their needs. When I compressed wise use principles in a paragraph, I realized, with distress, that they match the core beliefs of mainstream society: A free market system provides the greatest good for the greatest number of people. The natural world is not fragile but, rather, inexhaustible. Technology can meet any challenge, solve any problem. Growth equals progress, and, so, accumulation of material goods measures success. Individual liberty always trumps community rights. Humans stand at the center of the universe, and "man's" reworking of the earth is inevitably benevolent.

What I've always thought of as a fringe statement turns out to be a swatch from the fabric of America....

*Manifesto* is a fightin' word. Dogmatic by nature. Perhaps what I'm after is a credo: "This I believe." I want to take a stand based on introspection rather than a knee-jerk reaction, a stand that challenges dogma and articulates an ethic to live by as the middle-aged West fills in and rushes toward maturity.

Like our lives, my credo remains a work in progress. I offer it here now as a statement of what I believe today about land, community, and honor after fifty-odd years of living in the Imperial West and learning from the past, from my mentors, [*from the stories of impassioned advocates on all sides*], from the mountain, and from the mesa. A statement, too, of my dreams for the future.

My bulwark against regret.

### Credo: The People's West

*Lifelong locals know their home.* They understand the land's intimate cycles from decades and generations of living in place, a miracle of stability and identity.

*We can never hope to restore or sustain landscapes and watersheds without the cooperation of local citizens.* They rightfully resent and subvert any management scheme that excludes them from decision making. Collaboration is key.

*We need mutual trust, respect, empathy, and accountability.* The hits and misses of long-term elders can teach us all, while passionate newcomers—community members by choice—brandish a fierce love for their new home that can reinspire old-timers. Honor every skill and talent in the community. Having too many people at the table is always better than leaving someone out.

*Economic health is essential for community health.* If we don't create affordable housing and decent jobs for full-time residents, the community will lose its multigenerational roots. The working rural landscape will collapse into parody.

*Ecological health is essential for community health.* Conserve land *for* the land, and good things will come to people and community, as well.

*Rapid, unplanned growth profits only the boomer, rewards only the developer, and will in the long run fail citizens and destroy their sense of place.* Leadership must come from within the community. A master plan is the key to the future for each landscape—an inclusive, place-specific vision for a just transition to the 21st century economy, conceived in the broadest possible dialogue.

*Proliferating roads and off-road-vehicle use fragment the integrity of surviving wildlands.* Concentrate development where it already exists. Preserve agricultural land and the wild habitats it holds.

*Ranching on public lands contributes to the American cultural quilt. But cows should have no special rights.* Where cattle and sheep damage the land, reduce grazing and manage for restoration, resilience, and adaptation to climate change.

*That public lands make up most of the rural West is a positive—an asset.* Keep public lands public to create a buffer between village and wildland. With privatization of the commons, we lose community access.

*Refuse to drown in the deluge of change. Channel those floodwaters to power community dialogue.* Continually reassess any plan for a specific landscape and its neighborhoods. Insist on ecological sustainability, health, preservation of cultural tradition, and protection of biodiversity. Keep talking, no matter what. Keep listening, no matter what. Restraint is both visionary and conservative.

*Wildness is everywhere, but wilderness is a special category.* Designate and preserve large wilderness areas on public land wherever possible—several in each bioregion and connected by corridors. Establish local and regional land trusts to purchase critical private lands and hold conservation easements.

*One person, one passionate person speaking out stubbornly and relentlessly, can still make a difference.* Hard work by one individual can start a revolution and trigger crucial community dialogue.

*Arrogance is the opposite of relationship. Don't hesitate to use words like compassion and love and honor.* Depoliticize and humanize the issues, and fling open the windows on bureaucracy and authority. Remove obstacles to healing.

We are stuck with our untidy web of conflicting values. We all have our Edens, our devils, our bargains to strike. We are responsible for planning, making decisions, acknowledging duty, accepting stewardship—and for wrangling through as a community.

Start the conversation before a crisis. Share information and frustrations and dreams and anger and joy. Stomp along the riverbank together. Work together. Cook and eat together, tell stories together. Laugh together. Thrash through conflict to higher ground. Inclusivity requires trust and openness from old-timers and newcomers alike.

We call it paradise, this land of ours. We call it home. Like our nation, the West is in the middle of its arc. We must remain both vigilant and tender if we wish to preserve its authenticity. We can do this. As a community, we are not yet too old, too greedy, or too cynical to take wise action together.

# CAPITOL REEF TIMELINE

12,000 years ago: Humans first live in Capitol Reef country.

2,500 years ago: Ancestral Puebloans and their predecessors, including the ancestors of the Zuni and Hopi people, document in rock art their passage through Capitol Reef.

800–1200: Peak population of Fremont Culture.

1300s–1900s: Ute and Paiute people include Capitol Reef in their territories.

1600s–1800s: Capitol Reef is a place of refuge for Navajo families.

1854: Solomon Carvalho photographs Cathedral Valley obelisks as John Charles Frement's expedition passes through the future national park during a winter storm.

1866: Mormon militia companies reach Capitol Reef. William Byron Pace follows the Fremont River to Fruita. James Andrus looks out from Boulder Mountain; his expedition diarist, Franklin Wooley, describes the view over the Waterpocket Fold.

1872: Almon Thompson leads Powell government survey team from Boulder Mountain through Pleasant Creek to the Henry Mountains; Jack Hillers photographs the route.

1870s: Mormon pioneers establish western Wayne County towns in Rabbit Valley.

1881–1884: Charles Hall operates Colorado River ferry at Halls Crossing as an alternative to Hole-in-the-Rock trail.

1882: Ephraim K. Hanks moves to Floral Ranch on Pleasant Creek.

1883–1884: Elijah Cutler Behunin family lives in a cabin along Fremont River; Behunin leads the party that clears the wagon road through Capitol Gorge.

1885: Franklin Young builds the first home in Fruita but leaves the next year. Nels Johnson builds the first permanent home in 1886.

1896: Fruita schoolhouse is constructed.

1897: Catastrophic annual floods on the Fremont River trigger erosion that leads to an entrenched stream and the abandonment of most villages downstream from Fruita by the early 1900s.

About 1900: The forty-six residents of the village at the junction of the Fremont River and Sulphur Creek gain a post office. They had been calling their community "Junction," but that name was already taken, so the town became "Fruita."

1925: Utah governor George Dern comes to Fruita (population, just over one hundred) to celebrate proposed Wayne Wonderland State Park. Park booster Joseph Hickman drowns in Fish Lake four days later.

1926: E. P. Pectol excavates three buffalo hide shields from a cave near Torrey.

1928–1929: Noel Morss conducts archaeological fieldwork in Wayne County that leads to his recognition of the Fremont people as a distinct culture.
1929: Last regional Paiute/Ute Sun Dance is held at Fish Lake.
1930s: Charles B. Hunt conducts geological fieldwork in the Henry Mountains by pack train.
1932: National Park Service official Roger Toll begins the first official investigation of proposed Wayne Wonderland National Park or Monument.
1937, August 2: President Franklin D. Roosevelt creates a 37,060-acre Capitol Reef National Monument.
1938–1942: Young men with the Civilian Conservation Corps camp below Chimney Rock and work on trails, bridges, dams, a ranger station, and roads.
1939: Lurt Knee purchases Floral Ranch/Sleeping Rainbow Guest Ranch from Levi and Billie Bullard.
1941: Last classes are taught in Fruita schoolhouse.
1943: After moving to Fruita in 1941, Charles Kelly begins his career with the National Park Service as the first unpaid Capitol Reef custodian—with free housing.
1945: Charles Kelly and Frank Beckwith name Cathedral Valley.
1947: Formerly a stock route pioneered by John Atlantic Burr, the Burr Trail becomes passable to four-wheel-drive vehicles.
1952: Charles Kelly becomes the first Capitol Reef superintendent; he retires in 1959.
1952–1956: Capitol Reef National Monument is open to uranium exploration.
1957: After Utah Highway 24 is paved from Torrey to Fruita, monument visitation jumps from 7,500 in 1956 to 62,500 in 1957.
1958: President Dwight Eisenhower enlarges Capitol Reef National Monument by 3,040 acres. These boundary adjustments extend the monument southward to the north rim of Sulphur Creek as well as embracing all of old Highway 24 (today's Scenic Drive).
1962: A new route for Utah Highway 24 is paved and opened along Fremont River canyon; the old Capitol Gorge road is permanently closed to through traffic.
1964: National Park Service completes purchase of all but two properties in Fruita and removes most structures on these properties by the end of the decade.
1969, January 20: Using boundaries proposed by Superintendent Robert Heyder, President Lyndon Johnson expands Capitol Reef National Monument by 600 percent on his last day in office.
1971, December 18: Congress establishes Capitol Reef National Park.
1989: Annual park visitation passes 500,000 for the first time.
1991: Burr Trail is paved between Boulder and Capitol Reef National Park boundary.
2003: Pectol shields are returned to Navajo Nation.
2008: Utah Valley University Capitol Reef Field Station opens in Pleasant Creek.
2015: Capitol Reef receives "Gold Tier" status as an International Dark Sky Park
2016: Annual park visitation passes one million for the first time.

# FURTHER READING


Abbey, Edward. *Desert Solitaire: A Season in the Wilderness.* New York: McGraw-Hill, 1968.

Abbey, Edward, and Philip Hyde. *Slickrock: The Canyon Country of Southeast Utah.* San Francisco: Sierra Club Books, 1971.

Allen, Steve. *Utah's Canyon Country Place Names.* Durango, CO: Canyon Country Press, 2012.

Fillmore, Robert. *Geological Evolution of the Colorado Plateau of Eastern Utah and Western Colorado.* Salt Lake City: University of Utah Press, 2011.

Fowler, Don D. *The Glen Canyon Country: A Personal Memoir.* Salt Lake City: University of Utah Press, 2011.

Hebner, William Logan. *Southern Paiute: A Portrait.* Photographs by Michael L. Plyler. Logan: Utah State University Press, 2010.

Janetski, Joel C., Lee Kreutzer, Richard K. Talbot, Lane D. Richens, and Shane A. Baker. *Life on the Edge: Archaeology in Capitol Reef National Park.* Museum of Peoples and Cultures, Occasional Papers No. 11. Provo, UT: Brigham Young University, 2005.

Kelly, Charles. *The Outlaw Trail: A History of Butch Cassidy and His Wild Bunch.* New York: Bonanza Books, 1959.

McEntire, Frank, ed. *Final Light: The Life and Art of V. Douglas Snow.* Salt Lake City: University of Utah Press, 2013.

Nabhan, Gary, and Caroline Wilson. *Canyons of Color: Utah's Slickrock Wildlands.* Genesis Series. New York: Tehabi Books and Harper Collins*West*, 1995.

Probasco, Christian. *Highway 12.* Logan: Utah State University Press, 2005.

Rogers, Jedediah. *Roads in the Wilderness: Conflict in Canyon Country.* Salt Lake City: University of Utah Press, 2013.

Roylance, Ward J. *The Enchanted Wilderness: A Red Rock Odyssey*, Torrey, UT: Four Corners West, 1986.

Simms, Steven R. *Traces of Fremont: Society and Rock Art in Ancient Utah.* Photographs by François Gohier. Salt Lake City: University of Utah Press, 2010.

Snow, Anne, comp. *Rainbow Views: A History of Wayne County.* 4th ed. Wayne County, UT: Daughters of the Utah Pioneers, 1985.

Stegner, Wallace. *Beyond the Hundredth Meridian: John Wesley Powell and the Second Opening of the West.* Boston: Houghton Mifflin, 1954.

Swanson, Frederick H. *Dave Rust: A Life in the Canyons.* Salt Lake City: University of Utah Press, 2007.

Trimble, Stephen. *The People: Indians of the American Southwest*. Santa Fe, NM: SAR Press, 1993.

———, ed. *Red Rock Stories: Three Generations of Writers Speak on Behalf of Utah's Public Lands*. Torrey, UT: Torrey House Press, 2017.

Wilkinson, Charles. *Fire on the Plateau: Conflict and Endurance in the American Southwest*. Washington, DC: Island Press, 2004.

Williams, Terry Tempest. *Red: Passion and Patience in the Desert*. New York: Pantheon Books/Random House, 2001.

# ACKNOWLEDGMENTS

Thanks to the following readers and scholars who critiqued all or part of the manuscript and provided valuable feedback on early drafts: Cathy Bagley, Ken Jones, Jana Richman, Jedediah Rogers, Robert Shlaer, Steve Simms, Chuck Smith, Judy Smith, Rosemary Sucec, Steve Taylor.

Thanks to these folks, who provided new material for this reader or helped with leads: Steve Allen, Adus Dorsey, Don Fowler, Steve Lutz, Kate MacLeod, Kiri Manookin, Frank McEntire, Jen Jackson Quintano, Barry Scholl, James Swensen, Guy Tal, Ann Whittaker.

The Capitol Reef National Park and Natural History Association staff graciously assisted with this project. Thanks to Sue Fritzke, Cindy Micheli, Lori Rome, Ed Fallis, and Shirley Torgerson. In the 1970s, I worked for two stellar NPS rangers at Capitol Reef, Chief Naturalist Jay Cable and Superintendent Bob Reynolds. This book harvests the fruits of their support and their enthusiasm for sending me into the backcountry to photograph. I first discovered the park on hikes with Michael Doubleday, Leslie Kathryn Foxworth, Bill Hauze, Dave and Peg Johnson, Lee Keyser, Linda Yale, and Mark Zarn. Years later, Fred Goodsell shared his unparalleled knowledge of Capitol Reef backcountry.

Thanks to my series editors and the production team at the University of Utah, especially John Alley and David Stanley. Dave provided a steady hand for the project and asked hard questions about crucial details. John enthusiastically embraced a big vision for the book. He suggested running color throughout, a luxury I hadn't considered. A generous grant from the The Nature Conservancy of Utah made this possible. Thanks to Dave Livermore, Utah state director of The Nature Conservancy, for his support of this book—and for his big-hearted championing of my work over many years.

When John Alley retired, Reba Rauch and Patrick Hadley gave the project their full attention. Laurel Anderton copyedited with sharp eyes. And Jessica Booth created an elegant design for a complicated project. Thanks to you all.

Lastly, thanks to my family—especially my wife, Joanne Slotnik, our kids, Dory Trimble and Jake Trimble, and our cousin, Carol Dayn. They have been my hiking buddies for thirty years and remarkably patient with the thousands of times I've hollered at them on the trail, "Please stop there for a picture!"

## SOURCES AND CHAPTER NOTES

### Editor's note

I've included bracketed notes here and there to clarify meaning or to add a comment. Throughout the reader I italicize my text within these brackets. If the original text included bracketed asides, they remain unitalicized.

### Frontispiece

Elijah Cutler Behunin led the 1883 work party to build a road through the Waterpocket Fold. This Capitol Gorge route remained the only passenger vehicle route across south-central Utah until 1962, when Highway 24 was paved through the Fremont River canyon. Inglesby photograph collection, Special Collections, J. Willard Marriott Library, University of Utah.

### Epigraph

Edward Abbey, "A Thirst for the Desert," in *Wilderness USA*, National Geographic Society, 1973.

### Introduction

Page 1: Based, in part, on my essay "The Blue Gate" in *Nature, Love, Medicine*, Thomas Lowe Fleischner, editor, Torrey House Press, 2017. Used with permission.

Page 4: Map by David Fuller, DLF Group, Santa Barbara, California, from George Davidson, *Red Rock Eden*, copyright © 1986 by Capitol Reef Natural History Association, Torrey, Utah. Edited and reprinted 2002. Used with permission.

Page 6: Geologic cross section from *Ancient Landscapes of the Colorado Plateau* by Ron Blakey and Wayne Ranney, Grand Canyon Association, Grand Canyon National Park, Arizona, 2008. Used with permission of Ron Blakey, Colorado Plateau Geosystems/Deep Time Maps.

### The View from Boulder Mountain

Page 17: Franklin B. Wooley 1866 journal entry, p. 156 in "Military Reconnaissance in Southern Utah, 1866," C. Gregory Crampton, editor, *Utah Historical Quarterly* 32, no. 2 (Spring 1964). Copyright © Utah Historical Society. Used with permission. (The original Wooley manuscript is in the Military Records Section of the Utah State Archives.)

Page 19: Clarence E. Dutton, *Report on the Geology of the High Plateaus of Utah*, US Geographical and Geological Survey of the Rocky Mountain Region, 1880.

Page 22: Wallace Stegner, "Coda: Wilderness Letter," in *The Sound of Mountain Water* by Wallace Stegner, copyright © 1946, 1947, 1949, 1950, 1952, 1958, 1959, 1961, 1963, 1965, 1966, 1967, 1969, 1980 by Wallace Stegner. Used by permission of Doubleday, an imprint of the Knopf Doubleday Publishing Group, a division of Penguin Random House LLC. All rights reserved.

Page 24: Linda Elizabeth Peterson, "Into the Escalante," in *What Wildness Is This: Women Write about the Southwest*, Susan Wittig Albert, Susan Hanson, Jan Epton Seale, Paula Stallings Yost, editors, copyright © 2007 by Story Circle Network, University of Texas Press, 2007. Used with permission of Susan Wittig Albert.

### Native Heritage

Page 35: David Madsen, *Exploring the Fremont*, copyright © 1989 by Utah Museum of Natural History, Salt Lake City, Utah. Used with permission.

Page 37: Dee Hatch interview. Used with permission of Steve Allen.

Page 39: Rose Houk, *Dwellers of the Rainbow: The Fremont Culture in Capitol Reef Country*, copyright © 1988 by Capitol Reef Natural History Association, Capitol Reef National Park, Torrey, Utah. Used with permission.

Page 44: Fremont cradleboard with figurine, photograph copyright © François Gohier. From Steven R. Simms, *Traces of Fremont: Society and Rock Art in Ancient Utah*, photographs by François Gohier, University of Utah Press, 2010. Used with permission.

Page 46: Steven R. Simms, *Ancient Peoples of the Great Basin and Colorado Plateau*, copyright © 2008 by Left Coast Press, Walnut Creek, California. Used with permission of the author.

Page 51: Rosemary Sucec, *Fulfilling Destinies, Sustaining Lives: The Landscape of the Waterpocket Fold (An Ethnographic Overview and Assessment of American Indian Histories and Resource Uses within Capitol Reef National Park, Utah and on Lands Surrounding It)*, National Park Service, Intermountain Region, Denver, Colorado, 2006. Used with permission.

Page 56: Robert S. McPherson, "Seeing Is Believing: The Odyssey of the Pectol Shields," in *Dinéjí Na`nitin: Navajo Traditional Teachings and History*, copyright © 2012 by University Press of Colorado, Boulder. Used with permission of the University Press of Colorado.

Page 58: Photograph of Ephraim Pectol and his two daughters. Used with permission of the Pectol Family Organization.

Page 62: John Holiday. Photograph by Gavin Noyes, Utah Diné Bikéyah. Used with permission.

### Exploration

Page 68: G.K. Gilbert photo. Department of the Interior/US Geological Survey, https://library.usgs.gov/photo/#/item/51dda23be4b0f72b4471df2d.

Page 68–69: Charles B. Hunt quotes from interview with author. Manuscript in Stephen Trimble papers, Accn 2757, Box 60, Special Collections and Archives, J. Willard Marriott Library, University of Utah, Salt Lake City.

Page 71: Solomon Nunes Carvalho, *Incidents of Travel and Adventure in the Far West: With Colonel Fremont's Last Expedition across the Rocky Mountains, Including Three Months' Residence in Utah, and a Perilous Trip across the Great American Desert to the Pacific*, New York: Derby & Jackson, 1858; reprint, University of Nebraska Press, Bison Books, 2004.

Page 79: Robert Shlaer, *Sights Once Seen: Daguerreotyping Fremont's Last Expedition through the Rockies*, copyright © 2000 by Museum of New Mexico Press, Albuquerque. Used with permission of the author.

Page 82: *Natural Obelisks*. Steel engraving from the prospectus for John Charles Fremont's *Memoirs of My Life*, based on a daguerreotype by Solomon Carvalho. Courtesy the Huntington Library, San Marino, California.

Page 83: *Natural Obelisks, "Mom, Pop, and Henry," Capitol Reef, 1996*. Daguerreotype by Robert Shlaer. Used with permission.

Page 85: Jack Hillers, "Photographed All the Best Scenery: Jack Hillers's Diary of the Powell Expeditions, 1871–1875," in *Cleaving an Unknown World: The Powell Expeditions and the Scientific Exploration of the Colorado Plateau*, Don D. Fowler, editor, copyright © 2012 by University of Utah Press. Used with permission.

Page 87: J.K. Hillers photo, *Old Time rocks. Tantalus Creek, Aquarius Plateau, Wayne County, Utah*. Courtesy National Archives, photo no. 57-PS-420.

Page 90: Charles B. Hunt, *Geology of the Henry Mountains, Utah, as Recorded in the Notebooks of G.K. Gilbert, 1875–76*, Geological Society of America Memoir 167, copyright © 1988 by Geological Society of America, Boulder, Colorado. Used with permission.

Page 94: George C. Fraser, *Journeys in the Canyon Lands of Utah and Arizona 1914–1916*, Frederick H. Swanson, editor, copyright © 2005 Arizona Board of Regents. Reprinted by permission of the University of Arizona Press.

Page 94: David Rust background and quote from William W. Slaughter, "Dave Rust: Pioneering Outdoorsman and Guide: A Force behind Making Utah's Desert Famous," *Entrada Journal*, online post, May 27, 2010, http://www.entradainstitute.org/2010/05/dave-rust-pioneeing-outdoorsman-and-guide/.

Page 99: John A. Widtsoe, "A Journal of John A. Widtsoe, Colorado River Party, September 3–19, 1922, Preliminary to the Santa Fe Conference Which Framed the Colorado River Compact," A.R. Mortensen, editor, *Utah Historical Quarterly* 22 (1955): 195–231, copyright © Utah Historical Society. Used with permission.

## Canyon Country

Page 109: Ward J. Roylance, *Seeing Capitol Reef National Park: A Guide to the Roads and Trails*, copyright © 1979 by Wasatch Publishers, Salt Lake City, Utah. The Ward Roylance papers can be found in Accn 1282, Special Collections and Archives, J. Willard Marriott Library, University of Utah, Salt Lake City.

Page 112: Edward Abbey, "Come on In," foreword to *Blessed by Light: Visions of the Colorado Plateau*, Stephen Trimble, editor, copyright © 1986 by Gibbs Smith Inc., Peregrine Smith Books, published by Gibbs M. Smith Inc., Layton, Utah. Used with permission of Clarke Abbey.

Page 117: Charles Bowden, *Stone Canyons of the Colorado Plateau*, photographs by Jack Dykinga, text by Charles Bowden, text copyright © 1996 by Charles Bowden, Harry N. Abrams, New York. Used with permission of the Charles Clyde Bowden Literary Trust, Mary-Martha Miles, trustee.

Page 122: Michael Collier, "The Neverlasting Hills," in *The Geology of Capitol Reef National Park*, copyright © 1987 by the Capitol Reef Natural History Association, Torrey, Utah. Used with permission of the author. Quote in headnote from a 2007 NPR interview, http://www.npr.org/2007/05/28/10408149/birds-eye-photographer-shoots-from-his-cessna.

Page 128: Ellen Meloy, "A Field Guide to Brazen Harlotry," in *The Anthropology of Turquoise: Meditations on Landscape, Art, and Spirit* by Ellen Meloy, copyright © 2002 by Ellen Meloy. Used by permission of Pantheon Books, an imprint of the Knopf Doubleday Publishing Group, a division of Penguin Random House LLC. All rights reserved.

Page 135: Ann Zwinger, *Aspen: Blazon of the High Country*, text by Ann Zwinger, photographs by Barbara Sparks, text copyright © 1991 by Ann Zwinger, Gibbs Smith, Layton, Utah. Used with permission of the Ann Zwinger Archive, Tutt Library, Colorado College, Colorado Springs, Colorado.

Page 137: Ann Zwinger, *Wind in the Rock*, copyright © 1978 by Ann H. Zwinger, Harper & Row (1st ed.); University of Arizona Press paperback reprint. Used with permission of the Ann Zwinger Archive, Tutt Library, Colorado College, Colorado Springs, Colorado.

Page 139: Ronald M. Lanner, *The Piñon Pine: A Natural and Cultural History*, copyright © 1981 by Ronald M. Lanner, University of Nevada Press, Reno. Used with permission.

## Home

Page 147: Ruby Noyes Tippets, *A Song in Her Heart*, privately printed, 1962. Used with permission of Don D. Fowler.

Page 156: Lenard E. Brown, *Capitol Reef Historical Survey and Base Map*, National Park Service Division of History, Office of Archeology and Historic Preservation, June 30, 1969.

Page 159: Sidney A. Hanks and Ephraim K. Hanks Jr., *Scouting for the Mormons on the Great Frontier*, Deseret News Press, Salt Lake City, 1948.

Page 167: Billie Bullard oral history, from Bradford Frye interview, 1991, in Capitol Reef National Park archives; and the author's 1994 interview, archived in Stephen Trimble papers, Accn 2757, Special Collections and Archives, J. Willard Marriott Library, University of Utah, Salt Lake City. Used with permission.

Page 173: Lurt Knee oral history, from Bradford Frye interview, 1991, in Capitol Reef National Park archives; and the author's 1994 interview, archived in Stephen Trimble papers, Accn 2757, Special Collections and Archives, J. Willard Marriott Library, University of Utah, Salt Lake City. Used with permission.

Page 178: Chip Ward, "Landing under a Sleeping Rainbow," based on *Canaries on the Rim: Living Downwind in the West*, copyright © 1999 by Chip Ward, Verso Books, New York and London, 1999. Used with permission of the author.

Page 186: Thanks to Kiri Manookin for providing these charming Capitol Reef Field Station journal entries from her Utah Valley University students, copyright © 2017 by the authors, Kadidia Sabou Doumbia, Salma Hakeem, and Tetiana Novikova. Used with permission.

Page 191: Ray Conrad, "The O-Bar Drive," in *Fence Lines: Poems about Cowboys & Country Folk, Kid Groups and Congress, and Other Rhymes for Our Place and Times*, copyright © 2009 by Ray Conrad, Avalanche Creek Productions, Teasdale, Utah. Used with permission of the author.

### Fruita

Page 197: Wayne County historian Steve Taylor alerted me to William Byron Pace's 1866 exploration along the Fremont River as far as Fruita—a month before Franklin Wooley looked out over the Waterpocket Fold from Boulder Mountain with the James Andrus company of militia. For details, see John Alton Peterson, *Utah's Black Hawk War*, University of Utah Press, 1998, 311–12.

Page 199: *Pick, Pick Apples*, copyright © 2015, Kate MacLeod, Courier Music, ASCAP/ www.katemacleod.com. Used with permission. Music video at https://www.katemacleod.com/music-page.

Page 201: Franklin Wheeler Young, *My Dear Daughter: Lost Letters of Franklin Wheeler Young*, Ronald Bodtcher, editor, copyright © 2014 by Ronald Bodtcher. Wandering Press, North Hollywood, California. Used with permission.

Page 202: Clay M. Robinson, "Echoes of Childhood in Old Fruita," in *Echoes from the Cliffs of Capitol Reef National Park*, reminiscences from Max E. Robinson and Clay M. Robinson, copyright © 2004 by Clay M. Robinson, published by Ol' Gran'Pa Stories, West Jordan, Utah. Used with permission of the Robinson family.

Page 208: George Davidson, *Red Rock Eden*, copyright © 1986 by Capitol Reef Natural History Association. Edited and reprinted 2002. Used with permission.

Page 209: Fruita in the early days. Arthur Inglesby photo collection, P0899 n159, Special Collections, J. Willard Marriott Library, University of Utah, Salt Lake City.

Page 215: Jen Jackson Quintano, "Jeweled Jars of Memory," adapted from *Mountain Gazette* blog, September 12, 2011, http://www.mountaingazette.com/blogs/desert-reflections/jeweled-jars-of-memory/. Used with permission of the author.

Page 219: Wallace Stegner, "High Plateaus," in *American Places* by Eliot Porter, Wallace Stegner, and Page Stegner, edited by John Macrae III, copyright © 1981 by E.P. Dutton. Used by permission of Dutton, an imprint of Penguin Publishing Group, a division of Penguin Random House LLC. All rights reserved.

Page 221: Doc Inglesby photo. Utah State Historical Society Classified Photo Collection, number 11060. Used with permission.

Page 225: Renny Russell, *Rock Me on the Water: A Life on the Loose*, copyright © 2007 by Renny Russell, Animist Press, Questa, New Mexico. Used with permission.
Page 232: Gary Topping, "Charles Kelly's Glen Canyon Ventures and Adventures," in *Utah Historical Quarterly* 55, no. 2 (Spring 1987): 120–36, copyright © Utah Historical Society. Used with permission.
Page 233: Photo of Charles and Harriette Kelly. Charles Kelly Collection P0100n1_29_18, Special Collections, J. Willard Marriott Library, University of Utah, Salt Lake City.
Page 234: Jonathan Scott Thow, "Capitol Reef: The Forgotten National Park," master's thesis in history, Utah State University, Logan, 1986. Used with permission of the author.

## National Vision

Pages 241 and 267: Jared Farmer, *Glen Canyon Dammed*, © 1999 Arizona Board of Regents. Reprinted by permission of the University of Arizona Press.
Page 245: Monument dedication photo. Photo by A.L. Inglesby, Utah State Historical Society Classified Photo Collection, number 11165. Used with permission.
Page 247: Jonathan Scott Thow, "Capitol Reef: The Forgotten National Park," master's thesis in history, Utah State University, Logan, 1986. Used with permission of the author.
Page 248: US Geological Survey geologist R.Q. Lewis at the Oyler Mine, Grand Wash, 1952, photograph by E.N. Hinrichs, Department of the Interior/US Geological Survey, https://library.usgs.gov/photo/#/item/51dc89dce4b097e4d38397b5.
Page 254: Bradford J. Frye, *From Barrier to Crossroads: An Administrative History of Capitol Reef National Park, Utah*, Cultural Resources Selections No. 12, Intermountain Region, National Park Service, Denver, Colorado, 1998, https://www.nps.gov/park history/online_books/care/adhi/adhi.htm.
Page 260: Map of evolving park boundaries from Frye, *From Barrier to Crossroads*, fig. 30.
Page 265: Lurt Knee oral history, from Bradford Frye interview, 1991, in Capitol Reef National Park archives; and the author's 1994 interview, archived in Stephen Trimble papers, Accn 2757, Special Collections and Archives, J. Willard Marriott Library, University of Utah, Salt Lake City. Used with permission.

## Refuge

Page 277: James Swensen, "Capitol Reef's Most Famous Photograph," copyright © 2016 by James Swensen. This original essay was taken and expanded from a lecture on the history of photographing Capitol Reef National Park, Entrada Institute, Torrey, Utah, September 2, 2012. Used with permission.

The quotations used in the essay were taken from conversations with Drid Williams and Arthur Lazar as well as the following sources: Minor White, "Equivalence: The Perennial Trend," in *Photographers on Photography*, Nathan Lyons, editor (Englewood Cliffs, NJ: Prentice-Hall, 1966); James Baker Hall, *Minor White: Rites and Passages* (New York: Aperture, 1978); and *Minor White: A Living Remembrance* (New York: Aperture, 1984), which also features Robert Adams, "The Spring-Tight Line," and Drid Williams, "Creating the Space."

Page 280: Photograph of Minor White by Arthur Lazar. Used with permission.

Page 281: Minor White, American, 1908–1976, *Moenkopi Strata, Capitol Reef, Utah, 1962*. Gelatin silver print, image: 30.4 × 23.8 cm; sheet: 35.4 × 27.7 cm. Minor White Archive, Princeton University Art Museum, bequest of Minor White (x1980–3918), copyright © Trustees of Princeton University. Used with permission.

Page 284: Capitol Reef, Illustrated, based in part on my essays "Participating in Home: Following Wallace Stegner into the Heart of the West," in *Isotope: A Journal of Literary Nature and Science Writing*, Utah State University, Spring/Summer 2009; and "The Naming of Hyde's Wall," *Landscape Photography Blogger*, August 2012, http://landscapephotographyblogger.com/the-naming-of-hydes-wall-by-writer-and-photographer-stephen-trimble/.

Page 285: Bruce Barnbaum, *Boulder and Metamorphosis Wave, Capitol Reef, 1987/1990*, copyright © Bruce Barnbaum. Used with permission.

Page 286: John Pfahl, *Moonrise over Pie Pan, Capitol Reef National Park, Utah, 1977*. Dye imbibition print, image size: 20 × 25.6 cm (7⅞ × 10$\frac{1}{16}$ in.); sheet: 21.6 × 27.9 cm (8½ × 11 in.), copyright © John Pfahl, J. Paul Getty Museum, Los Angeles. Used with permission.

Page 287: Philip Hyde, *Canyon in Waterpocket Fold, Capitol Reef National Park, Utah, 1970*, copyright © Philip Hyde Photography. Used by permission of the Estate of Philip Hyde.

Page 288: Eliot Porter (1901–1990), *Pool and Willows, Waterpocket Fold, Utah, August 21, 1963*, dye imbibition print, 10$\frac{5}{16}$ × 8 in. Amon Carter Museum of American Art, Fort Worth, Texas, bequest of the artist, P1990.51.5208, © 1990 Amon Carter Museum of American Art. Used with permission.

Page 289: Guy Tal photograph, copyright © Guy Tal Photography. Used with permission.

Page 291: Doug Snow quotes and background from *Final Light: The Life and Art of V. Douglas Snow*, Frank McEntire, editor, University of Utah Press, 2013. Thanks to Frank McEntire for sharing the script of Snow's unpublished play "Blind Sight."

Page 292: Gary Ferguson, *Shouting at the Sky*, copyright © 1999 by Gary Ferguson, Thomas Dunne Books, St. Martin's Press, New York. Used with permission of the author. Quotes in headnote from a 2008 keynote lecture by Ferguson, http://www.legacyoutdooradventures.com/blog/2013/02/ten-years-after-gary-ferguson-interviews-former-wilderness-clients/.

Page 298: Ann Whittaker, "So What If We All Want Redemption?" copyright © 2016 by Ann Whittaker. Used with permission of the author.

Page 300: Greg Gordon, "Unconformity," in *Landscape of Desire: Identity and Nature in Utah's Canyon Country*, copyright © 2003 Utah State University Press, Logan. Used with permission of the University Press of Colorado.

Page 306: Stephen Trimble, *Rock Glow, Sky Shine: The Spirit of Capitol Reef*, copyright © 1978 Capitol Reef Natural History Association, Torrey, Utah. Used with permission.

Page 310: Rose Houk, "October Journey," in *Capitol Reef: Canyon Country Eden*, copyright © 1996 Capitol Reef Natural History Association, Torrey, Utah. Used with permission.

Page 316: Craig Childs, "The Trouble with Cellphones," blog post, "Writings: Places," October 4, 2008, http://www.houseofrain.com/rantsandwritings.cfm?mode=detail&id=1223127663206&group=writings. Used with permission of the author. Quotes in headnote from his website bio at www.houseofrain.com.

Page 318: Steve Howe, "National Parks: Capitol Reef. Discover a Private Paradise by Crossing the Park's Convoluted Waterpocket Fold," *Backpacker* magazine, June 2011. Used with permission of the author, http://www.backpacker.com/trips/utah/capitol-reef-national-park/national-parks-capitol-reef/#bp=0/img1. Headnote quotes from http://redrockadventureguides.com/about, where Steve is lead guide.

Page 323: Charles Riddel, *Spoked Dreams: An Odyssey by Bicycle and Mind*, copyright © 2005 by Charles S. Riddel, Air-Space Press, Austin, Texas. Used with permission of the author.

## Practicable Ridges

Page 331: Stephen Trimble, *Bargaining for Eden: The Fight for the Last Open Spaces in America*, copyright © 2008 Stephen Trimble, University of California Press, Berkeley and Los Angeles. Used with permission.

Page 332: Wheeler map, A. Humphreys, George M. Wheeler, Israel C. Russell, and United States Army Corps of Engineers, 1884. *Geological Atlas, Projected to Illustrate Geographical Explorations and Surveys West of the 100th Meridian of Longitude*. Southern and Southwestern Utah, Atlas Sheet 59. Courtesy of Special Collections, J. Willard Marriott Library, University of Utah, Salt Lake City.